RHETORICAL MEMORY

RHETORICAL MEMORY

A Study of Technical Communication
and Information Management

STEWART WHITTEMORE

THE UNIVERSITY OF CHICAGO PRESS

CHICAGO AND LONDON

STEWART WHITTEMORE is associate professor of English at Auburn University.

The University of Chicago Press, Chicago 60637
The University of Chicago Press, Ltd., London
© 2015 by The University of Chicago
All rights reserved. Published 2015.
Printed in the United States of America

24 23 22 21 20 19 18 17 16 15   1 2 3 4 5

ISBN-13: 978-0-226-26338-0 (cloth)
ISBN-13: 978-0-226-26341-0 (e-book)
DOI: 10.7208/chicago/9780226263410.001.0001

Library of Congress Cataloging-in-Publication Data
Whittemore, Stewart, author.
    Rhetorical memory : a study of technical communication and information
management / Stewart Whittemore.
        pages ; cm
    Includes bibliographical references and index.
        ISBN 978-0-226-26338-0 (cloth : alk. paper) — ISBN 978-0-226-26341-0
(ebook) 1. Communication of technical information—United States—Case
studies. 2. Knowledge management—United States—Case studies. 3. Memory.
4. Communication of technical information. 5. Knowledge management.
6. Software architecture. I. Title.
        T10.5.W49 2015
        658.4'038—dc23

                                                                2014048488

FOR ERIN

CONTENTS

ACKNOWLEDGMENTS

A book is a black box concealing the labor of many hands, not just that of the person whose name appears on the cover. So I'd like to thank the many people whose selfless labors on my behalf have made this book possible.

This project began as my dissertation at Michigan State University, so I'll begin by thanking my research participants, for sharing their time and considerable talents with me, and the members of my committee, Bill Hart-Davidson, Jim Porter, and Julie Lindquist, for their insights and advice, which led me to believe that there was a book. Most of all, of course, my eternal thanks goes to my chair, Jeff Grabill, whose constant cheerleading and occasional scolding kept this project alive through many years and over many rough patches.

Second, I'd like to thank the staff at the University of Chicago Press, especially my editor David Morrow and the anonymous reviewers whose guidance through multiple rounds of revision have made this a far stronger book. I also offer my sincere thanks to my copyeditor, Erik Carlson, for his keen eye and wry commentary, which made the process of whipping my flabby prose into shape quite enjoyable.

I would like to offer a special thanks to my first editor, Kirk St. Amant, for his faith in this project. Similarly, I owe a great deal of thanks to my colleagues at Auburn University, especially to Isabelle Thompson and Heath Fowler, for their moral support, encouragement, and advice.

My final words of thanks go out to my family—Erin, Owen, Andrew, Kat, Lavinia, and my mother and father—whose support and love have seen it through.

Managing Information in Organizations: An Overview of Ideas about Memory and Memory Work

Memory work is also process, like a journey; it must therefore have a starting-point.
—Mary Carruthers, 1998, p. 23

In the book *In the Key of Genius: The Extraordinary Life of Derek Paravicini*, professor of music Adam Ockelford tells the inspiring story of a blind and severely autistic young man who, despite seemingly impossible odds, has become a famous virtuoso pianist. Rather than proving to be a handicap, Derek's Paravicini's condition enables him to do something amazing. Only a short time after he listens to a piece of music—any piece of music, no matter how complex—he is able to replay it perfectly from memory, like a human Dictaphone. But Derek is more than a mere recorder. He is also able to improvise, to riff, and to embellish the music, making him a virtuoso at his art rather than some sort of "human iPod," able only to imitate, not to innovate or to improve upon the original composition (Ockelford, 2007, p. 45). Derek is able to do this in part, Ockelford explains, because his perfect embodied knowledge of his workspace—his piano keyboard—unites with his innate perfect pitch and aural memory to create a single psychosomatic (mind-body) memory. So firm is this embodied memory that recall is automatic and virtually effortless, which leaves Derek's mind free to weave amazing improvisations around the original material. Consequently, Derek's memory ensures that, despite what might at first seem to be a disability, he has achieved a successful and fulfilling career.

In our own careers, most of us are not, of course, able to perform our jobs with as much facility as Derek. Our brains are not wired to give us Derek's intense focus or perfect memory for detail. Similarly, our workplaces do not, by and large, mesh as seamlessly with our work activities as Derek's piano does with his. Consequently we lose things. We forget key pieces of information. We become distracted easily. Our in-boxes accumulate mail, our cell phones ring, our instant messaging icons flash, our list of to-dos grows longer, and the piles of paper on our desks grow taller. Together these distractions can cause information overload, the condition popularized by the futurist Alvin Toffler in the 1970s, in which completing even simple tasks seems to require a wearying amount of sifting and searching for relevant data. Our working lives thus become a delicate balancing act of trying to stay on top of the information that comes in as a routine part of our jobs while also exercising some creativity, of putting our own touch on this information in order to create value for ourselves and our employers. And, despite the explosion of revolutionary new technological prostheses that are supposed to make managing information in our lives and work easier, we rarely encounter anyone who feels firmly in control of his or her information workspace. Simply consider the variety of devices with which it is now possible to derive information: personal computers, televisions, personal data assistants (PDAs), iPods, tablets, smartphones, e-book readers, and even network-enabled smart wristwatches. And these are just some of the hardware platforms; it would be impracticable to list the multitude of software and network technologies that run on each device or the multiple human stakeholders who must be consulted in virtually any business process in a networked global economy. With all due respect and admiration for the exquisite talents of Derek Paravicini, the average office worker is bombarded with far more cognitive, sensory, and physical information inputs than Derek must deal with. And the result is that, with such a profusion of inputs from this wide array of sources, it can become difficult if not impossible to turn our attention to pure innovation as Derek does. Yet we know that such innovation and creativity are often essential to solving complex problems or finding new and better approaches to addressing issues.

Those of us who earn our living as writers in organizations, whether in industry or the academy, understand all too well the impediment to creativity and original, inventive thinking posed by the need to stay on top of incoming information—impediments that, ironically, appear greater the more information technology the employer utilizes for our supposed benefit, so that technical communicators in high-technology firms face perhaps the greatest risk of being overwhelmed by information of any of us. For example, in the research study I will be discussing throughout this book, which studied a team of technical communicators at the software firm Software Unlimited,[1] when asked to describe the principal difficulties entailed in managing projects, each of the three principal research participants, whose stories ground the case studies in chapters 4–6, focused on separate but related challenges of managing, storing, and using incoming information. Three telling quotations from these discussions offer a preview of the problems that will be discussed in succeeding chapters.

Consider first the trifecta of difficulties that Robert, the newest technical communicator at Software Unlimited, whose brief tenure with the firm will be considered in detail in chapter 4 in order to raise issues of information and memory affecting newcomers to a firm, identified when asked about the work he had been doing on his first project with the firm:

> We had these three barriers that made it hard. Number 1, the software to create the help was only on Windows and therefore I had to get files from here, so that's one area that's annoying. Number 2, all the changing screenshots. So basically, you are replacing a screenshot or maybe adding a new one with a file that's here, and so you better make sure that the old one's gone. I try really hard to save it as the same name and just overwrite it. And the third thing is you have the whole Windows/Mac thing. So, trying to take a document that's 90 percent the same for each situation but you might have to go through and change the task bars in all the doc. So you had two near-identical versions and then bad things can happen.

Second, Lucy, whose recent return to full-time work after family leave will give us an opportunity to consider issues of memory and information facing working parents in chapter 5, characterizes the principal difficulties of managing projects in terms of scheduling and team dependences:

> With our group, our projects—our tasks—don't relate. . . . It's not always easy to see what they relate to because we [i.e., the technical writers] might have something like "create PowerPoint" just to give an example, but below that heading are three tasks for three [software developers] to do. . . . So, I'll have the PowerPoint documentation but I need all those other things done before I can do anything with it. . . . I'm having [impediments] because things aren't getting done. Because I don't know when things should be done, you know, so what I'm trying to do is track when the things should be done. Sometimes something is supposed to be done and no one has even mentioned that they've worked on it or are going to work on it or anything because we are missing something with this tracking.

And third, when asked to describe the methods and tools that Software Unlimited provides to help her track her project deliverables—finalized written products like user manuals or online help files—Angela, the most senior technical communicator, understandably chooses to discuss the issue in terms of growing and changing organizational dynamics:

> This [i.e., tracking project deliverables] has always been a challenge. Now that we have six people on the documentation team—before it was just me; now there are six. We need standards. We need to be sure everybody is aware of the style guide and the things that we come up with. . . . We also have a training department that we interface with. So now we have all of their information and our combined information that we have to keep track of. And now we've got processes that are going into position for posting information on the Web. . . . There are a lot of meetings. We have a lot of projects and projects within projects. A release isn't just [the online help and regular documentation] anymore. There's the localization part. There's the web documentation part. There's the web marketing part. There's the training part. So we have lots of projects within major release projects. There's not a great way of keeping track of all that information at a glance, *so you really rely on remembering a lot.*

The difficulty of juggling multiple and overlapping versions of deliverables, the risk of miscommunicating critical information among teams with varying skillsets and work goals, and the sheer proliferation of interested parties involved in most design efforts—Angela and her colleagues' descriptions of conditions at her workplace probably sound all too familiar. The growing complexity of organizations, work teams, business processes, and technologies places more and more demand on our limited ability to pay attention, to remember, and to think. Like Angela and her fellow employees, we must rely upon our own limited resources of time, attention, and memory just to stay on top of this glut of information.

Software Unlimited is, in other words, typical of the workplaces in which many professional and technical communicators find themselves, and the issues Angela raises of managing information with only our own fallible human memories to guide us are ones that many of us can recognize and relate to. Fortunately for us and for Angela and her colleagues, most of our jobs do not require a perfect memory. What our jobs do require, however, is that we be able to locate the information we need when we need it and that we then analyze, mix, translate, sort, filter, and repurpose this information in order to meet the demands of the marketplace, to meet user needs, to satisfy customers, to feed upstream business processes, or to help ourselves and our colleagues make decisions. In addition—and this is a critical point—like Angela, our jobs require that we do this demanding work within specific physical, technological, and social spaces, which present us with unique sets of affordances (i.e., the perceivable properties of an object that indicate its possible uses) and constraints for managing information.

In short, technical communicators' relationship to information in their workplaces is rarely straightforward, and, in the absence of a perfect memory, mastering information requires a strategy. In this book, I use an examination of a particular case to help us better understand the issues confronting us in order to develop strategies for mastering information in our workplaces. To do so, I explore the experiences of Angela and her colleagues at Software Unlimited as they attempt both to stay on top of new, incoming, and shifting internal information for themselves and their teammates and to exercise creativity and add value to the organization's product offerings.

My intention is not to suggest that the strategies employed by the documentation team at Software Unlimited are necessarily the best ones, nor do I mean to imply that they have universal applicability to other workplaces. However, the information culture at Software Unlimited presents a rich environment for examining the challenges faced by technical communicators today. Specifically, like most software companies, Software Unlimited operates in a realm in which new information, usually in the form of technological advancement and innovation, is rapidly and constantly being created, both within and outside the company. It is critical for the company, and in particular its technical information managers, to be able to assimilate and manage new information, and especially to communicate and translate that information into the technical knowledge systems of its employees as well as the company as a whole, for the benefit of its customers.

At the same time and as a direct result of this influx of new and everchanging information, older technical information rapidly becomes outdated, obsolete, and superseded, requiring, in effect, that outmoded components of the technical knowledge existing within the company's information systems be purged to accord with such things as software updates and other improvements to the company's products. In other words, it is often just as important that old information be removed and discarded from the company's knowledge systems as it is that new information be assimilated into them.

Examining the work processes of the technical communicators at Software Unlimited, therefore, offers fertile ground for gaining a broad and insightful perspective on the problems faced by technical communicators and more importantly for discovering and evaluating the solution strategies that are developed by these communicators in the course of their daily operations. To gain such a perspective, we must first perform research. This book is about achieving a strategic perspective through such research.

SOME CAUSES OF INFORMATION OVERLOAD

A first step in addressing a problem is to define it by establishing a language for talking about it. So, before beginning a detailed examination of the experiences of the technical communicators at Software Unlimited, it is useful to consider why information is becoming such a burden in our workplaces. The ever-expanding number of information and communication technologies available to us is a good starting point for finding a common language. What is it about these technologies—many of which offer elegant user interfaces and useful tools for acquiring, sorting, displaying, and storing data—that paradoxically makes staying on top of information seem more, not less, difficult? To begin answering this question, we must first take a step back from the technologies themselves in order to better understand issues of information storage and use at a conceptual level.

One way of conceptualizing information inputs in our workplaces is to consider them in terms of *pushed* and *pulled* information (Kirsh, 2000). According to Kirsh, pushed information is data that arrive in our workspace (real or virtual) unasked for. It is sent to us without any prior request for it, as when a colleague forwards an email without first asking if we'd like to receive it. As recently as twenty years ago, this push-based information flow, though substantial, came in a rather limited range of media: paper memos and (physical) mail, newspapers, telephone calls, and so on. The range of tasks required to handle and manage this stream of data was similarly limited, as research has pointed out. For example, you could transcribe a spoken message (e.g., Anderson, 2004), enter an event in your calendar (e.g., Geisler, 2001), file a document in a filing cabinet (e.g., Yates, 1989), add a document to the pile on your desk (e.g., Malone, 1983), or, beginning in the early 1980s, stick a reminder to some surface with a sticky note (e.g., Spinuzzi, 2003b).

Now, of course, even though the stream of information arriving via paper documents has abated little despite the supposed dawn of the paperless office (e.g., Sellen & Harper, 2002), we must also figure out how to manage a never-ceasing, twenty-four-hour-a-day influx of instant messages, cell phone text messages, wiki updates, blog posts, tweets, Facebook updates, RSS feeds, and email on multiple mail accounts. Email alone is capable of geometrically expanding the amount of pushed information through the ability to add attachments of large (and multiple) documents and databases. (If we want to take this even further, attached documents or published articles can include links to other pertinent documents and articles—and so on—making the pool of available information appear infinite!) Pushed information, not surprisingly, has become a major source of stress among office workers (Edmunds & Morris, 2000). Wurman (2001) aptly labels this stress "information anxiety," because the burden of managing work information now spills over into our private lives, and all times of the day and night. Whereas, previously, pushed information arrived largely with the morning mail or through telephone contact during more or less specific "business hours," thanks to modern electronic communications (and in part to the burgeoning global economy), it is now capable of arriving 24/7, 365 days a year. As a result of our technologies, we now worry about what information-related deluge from work will take place while we are "off the clock." For example, will our in-boxes overflow if we are out sick? How will we ever find time to wade through our messages when we return from vacation? And, most pressingly, how do we do anything new when we spend all our time just trying to stay on top of our current information?

By contrast, pulled information is what we accumulate when we go looking for specific data, ideas, or opinions. It is what results when we purposefully search for answers to a specific question or browse about a general topic. It is data that do not come to us except by our own invitation. Until access to wide-area network technology and the World Wide Web began to become common in workplaces in the 1990s, our options for accumulating this pull-based information were also rather limited. We could, for example, try to recollect a piece of information from our own organic memory, consult our desktop dictionary, read the internal documentation, ask a coworker, interview a subject matter expert, search the files on our desktop computer, or look in our own or our organization's file cabinets. If we didn't have any luck with these options, some workplaces supplied encyclopedias or in-house libraries composed of sources related to our organization's core business. In a few cases, employees might have needed to go outside the organization to conduct research in public or university libraries and archives that held specific kinds of reference works. By and large, though, it was not too difficult to determine at what point we had done enough research (i.e., pull-based information gathering) to satisfy our employer or industry's need for sufficient due diligence.

While there are probably few technical communicators who would trade the affordances for research offered by the Internet with the far more limited ones available to us earlier, Internet technology nevertheless makes vastly

When did we cease before? (margin note)

more information accessible to discover, wade through, digest, and use. In so doing, the Internet broadened the base of our information-related job responsibilities. Kirsh (2000, p. 24) notes the irony of a situation in which "our life ought to get easier" and yet "no matter what we have found so far, most people harbor a lingering belief that even more relevant information lies outside, somewhere, and if found will save having to duplicate effort." Out of this information-rich context, new central questions emerged. How do we sort through it all? How do we know when we have consulted all the right sources? When have we pulled all the relevant information? When have we searched all the right databases? Indeed, all of these fears would seem to be well founded: The website MajesticSEO's Fresh Index claims that there are at least 200 billion web pages on the World Wide Web (which its search robots have actually visited) and estimates that there may be as many as 600 billion pages in total, although fewer than one-quarter of them are covered by search engines. It would seem that it is virtually impossible to "know" with anything close to absolute certainty that one has found the most precise and definitive information about anything. When do we cease looking, if only for the moment? And then of course, the question becomes, what do we do with the information once we have found it? How do we use it? How do we make it refindable? How do we communicate it to others? How do we convert raw information into usable knowledge and then make it accessible to coworkers or team members as well as to customers using the company's products? Finally, it is important to keep in mind that much of this information, both pushed and pulled, is editable; whereas the largely paper-based information of the past was static and unchangeable in its form, electronic documents very often are alterable by their users. In this way, the sheer volume of available information, coupled with both our increased access to it and our ability to manipulate that information, has led to significantly increased stresses related to managing it.

To further complicate matters, information obtained through the Internet can be repetitive or duplicated to varying degrees, and its reliability can be suspect. On the one hand, companies competing within specific industries are apt to "spin" either generic or technical information through online publications, blogs, or promotional materials as a means of increasing their visibility in the marketplace. On the other hand, the ability of just about anyone to be able to post information on the web—or of others to manipulate it—makes establishing the validity of web-based information a matter of concern for people wishing to use that information for their own business, educational, or other practical endeavors. It is certainly not unusual to find conflicting and/or repetitive data on the web, and there are whole websites devoted to debunking myths and hoaxes that nevertheless seem to rise like zombies year after year. Similarly, it seems to be the case that "bad" information—that is, information that is obsolete, outdated, or outmoded—is rarely systematically removed from the web, but simply lies out there in cyberspace, just like an old set of print encyclopedias from decades ago might lie on one's bookcase at home. The difference is, it's rather easy to know that those print encyclopedias collecting dust on the shelf are outdated; with respect to articles and documents posted on the web, that's not quite as easy to determine.

More pointedly, however, the result of this amassing of information via these multiple push and pull sources is a state of information fragmentation, a state in which the sheer abundance of information available often makes it impossible to see the big picture that our data are telling us. As Jones and Bruce (2005) put it, "Information that was once in paper form only is now scattered in multiple versions between paper and digital copies. Digital information is further scattered into 'information islands' each supported by a separate application or device" (p. 6). Such fragmentation compels us to interrupt work frequently in order to search for new information or to refind old information needed to accomplish job-related tasks, such as discovering the most telling statistic we need to finish writing a report, uncovering just the right piece of company lore we need to make a presentation persuasive, or correctly defining a piece of industry terminology to make ourselves appear credible to a client. And our work tools and environments, both digital and physical, often contribute to this sense of fragmentation by creating new interruptions and distractions that sidetrack our information-gathering and -processing objectives. For example, we start to work on a task, but a conversation erupts outside our cubicle wall, so we stop working in order to put on our headphones. Just then, our phone rings or an instant message pops up. Finally, our supervisor emails us with a task that needs to be completed ASAP, so we hastily try to preserve the state of the original work while establishing some context for undertaking this new task. The result is a combination of chaos, complexity, and frustration in which we spend an undue amount of time and energy just trying to stay on task.

INFORMATION OVERLOAD AND THE EVOLVING ROLE OF THE TECHNICAL COMMUNICATOR

Anxiety, fragmentation, duplication, interruption, loss of context—all of these take a toll on human attention, the "portal through which information reaches the brain," and attention is vital to our ability to retain information in memory (Klingberg, 2009, p. 19). Where Derek Paravicini's intimate and embodied familiarity with his workspace—principally his piano keyboard—reinforces his already astounding capacity for attention and memory, most technical communicators now find themselves in high-tech work environments that are far less conducive to sustained attention. Our workspaces are already fraught with all of the social distractions that have always existed since people first began to work in business offices.[2] And, all too often, each new technological innovation merely contributes to rather than alleviates the sensory overload and attention deficit that the knowledge worker experiences. Someone must take responsibility for paying attention to information, for managing it, and for making sure that it is not lost or forgotten when it is needed by the organization and its work teams. In short, someone needs to become the memory for the organization.

Increasingly, technical communicators are finding themselves filling this role of memory manager for their work teams and organizations. In some workplaces this development might be accidental—somebody had to do it, so it might as well be us. In many workplaces, however, this shift in responsibility may represent a natural evolution of job function long associated with writing. After all, as Ong (1982) and other literacy theorists argue, writing is *the* quintessential memory technology, and, consequently, those who write for their workplaces may be regarded as keepers of the common memory.

Whatever the immediate cause for such a shift, research indicates the functions and the duties of the technical communicator are rapidly changing in response to information. Within this context, a brief examination of this research will give us a better picture of what this evolving role for technical communicators entails. First, a growing body of field research points to the fact that technical communication work is becoming less about producing substantial written documents (e.g., user manuals or online help systems) and more about creating briefer and more ephemeral texts and verbal communications that can facilitate the free flow of information within organizations (Slattery, 2007; Hart & Conklin, 2006; Albers, 2005; Amidon & Blythe, 2008). Focus group studies, such as those of Wahl (2003) and Hart and Conklin (2006), found that technical communicators perceived traditional genres of documentation to be too static and inflexible to respond to the changing influx of new knowledge continually generated through software testing and interactions with customers. According to Hart and Conklin (2006, pp. 401–2), "Planning and facilitating communication processes" within organizations have become as large a part of technical communicators' responsibilities as producing end user documentation, if not a

larger part. Hart and Conklin speculate that the trend toward flattened hierarchies in organizations and more collaborative, team-focused work groups
in contemporary workplaces is driving this change. They further speculate
that these flattened arrangements mean that technical communicators are
becoming increasingly responsible not just for maintaining a one-way flow
of information from experts to users, but also for facilitating a two-way
flow of information throughout their teams and organizations. Albers (2005,
p. 269), moreover, notes that technical communicators who focus too narrowly on traditional documentation tasks at the expense of actively cooperating with their work teams to produce new approaches and products risk
"being relegated to the clerical position of taking notes and cleaning up the
team's reports."

Other studies of practicing technical communicators reinforce the conclusion that the new role of the technical communicator emphasizes bro-
kering information to multiple parties rather than focusing narrowly on
writing for a single audience. Amidon and Blythe (2008), for instance, find
that writing jobs that could be accomplished by writers working alone were
much more likely to be outsourced than those that required collaboration.
Jones (2005) reveals that some of these collaborative, information-brokering
activities include coordinating communications between divisions of the
company as well as maintaining the company intranet and knowledge base.
Similarly, Mott and Ford (2007) find that technical communicators are increasingly becoming information architects as they gain responsibilities for
managing information for their work teams.

Likewise, Conklin (2007) discovered a large shift in job responsibilities
over a five-year period as the technical communication group he studied
came to be perceived primarily as managers of organizational knowledge.
In particular, Conklin's participants noted that the collaborative work of
"facilitating information exchange and communication" had become a far
larger component of their jobs while activities associated with "writing and
editing texts" had remained flat by comparison (p. 217). Especially interesting, Conklin found that the technical communicators believe this work of
preserving, centralizing, integrating, and circulating information is critical
to their organization's ability to transform its stored information into useful knowledge. Conklin characterizes this job function as a creative process
of "knitting together" fragmented and isolated bits of information so that it
can be accessed and used most effectively (p. 227). Hovde (2001, p. 62) further emphasizes the inventive and creative aspects of this work: "Technical
communicators are not content to merely collect preexisting information
as if picking up apples that have fallen to the ground, but are involved in the

process of shaping meaning." In other words, in their new roles as information managers, technical communicators do not see themselves as merely regurgitating information like human iPods, but rather as creative shapers of information to meet the needs of diverse audiences. The findings by these researchers all seem to support the idea expressed earlier, that the explosion of information along with its ready accessibility from multiple sources has made it necessary for information managers not just to acquire, record, and circulate information, but also to synthesize and interpret that information into real, tangible knowledge in a form that is usable by others within the organization as well as by customers using the company's products.

In addition to adjusting to new team dynamics and new responsibilities, over the past ten years technical communicators have also witnessed the introduction of a variety of new technologies and methodologies for managing information. Most notable among these innovations were the introduction of content management systems and single-sourcing tools. Content management systems and single sourcing were, in fact, initially designed as solutions to the problem of information overload and as writing tools that would help technical communicators fulfill their newly evolving organizational roles. Single-sourcing software is intended to save writers from the burden of maintaining multiple versions of the same information. Text or other information can be authored once, entered in the database (the single source), and then reused and repurposed across documents. By comparison, the purpose of content management systems is to ensure that information does not become segregated in "content silos" on coworkers' hard drives or filing cabinets where other employees cannot find or use it (Rockley, Kostur, & Manning, 2003). To achieve this objective, content management systems employ markup technologies like XML to ensure that every piece of information available to an organization is accessible (i.e., findable) by any employee who needs it—especially by those engaged in writing tasks for the organization. Whether or not these tools are implemented or utilized effectively, their introduction into an organization contributes to the expectation that technical communicators become managers, brokers, and shapers of information across that organization (see Giammona, 2004; and Dicks, 2010, for further discussion of workplace developments affecting technical communicators in the first decades of the twenty-first century).

So, increasingly, the job of technical communication has begun to resemble that of information manager. Not only must we stay on top of our own information, pushed or pulled, but we must also maintain, facilitate, communicate, and otherwise transform information into useful and usable knowledge for our team members, our users, and a variety of other organizational players. This is a challenge indeed, and one that requires technical communicators to call upon their skills as both writers and as highly organized information managers. It also potentially raises the profile of technical communicators within their organization: they become the de facto go-to persons for issues of information sharing and availability throughout product design, development, rollout, and other work processes.

"Information Management"

PERSPECTIVES ON MEMORY, INFORMATION, AND KNOWLEDGE

Technical communicators are not, of course, the only professionals whose job descriptions are changing in response to developments in information technology. In fact, the technical communicator who goes looking for insights into how to adapt to and thrive in these changing circumstances is confronted by a vast literature on information, knowledge, and memory, each with roots in its own academic discipline, addressing its own priorities, and, occasionally, offering its own set of practical suggestions for addressing the problem. Each approach to memory offers a valuable perspective on the many ways in which human beings store, remember, and use past information to shape future actions. The approach to these issues that I adopt throughout this book builds on and references insights from four areas in particular. It will be helpful to review some of the highlights of the research in these areas in order to introduce the terminology and concepts that inform this book.

The first set of approaches that I draw from is cognitive approaches that study the operations of human memory in the brain or mind of the individual. These approaches arise from the fields of cognitive science and cognitive psychology. Memory researchers in cognitive science typically focus on the physiological processes by which the human brain captures, codes, stores, recollects, and forgets information (e.g., which neurons fire when a memory is triggered). Memory researchers in cognitive psychology similarly tend to focus on individual memory but concentrate more on describing and testing the psyche rather than the physical structures of the brain (e.g., the capacity and limitations of short- and long-term memory and the differences between procedural and declarative memory).

The approach to understanding memory work I present in this book owes a particular debt to two key insights identified by such studies of individual memory and cognition. First, beginning with the clinical research of Frederic Bartlett in the 1930s, studies of memory and cognition have shown that processes of human memory are fundamentally constructive and in-

Psychology

memory = experience

sensory

terpretive. In other words, human memory is not analogous to a computer database, in which retrieval is passive, a matter of "a simple table lookup" (Bannon & Kuutti, 1996, p. 161). Instead, remembering is an active process of reconstruction in which the circumstances and contexts in which a memory is originally stored (that is, the experience itself) often play a pivotal role in what we are able to do with the memory later. Second, also following Bartlett ([1932] 1964), cognitive research has demonstrated the particular durability of our memory for information when we perceive that information via multiple (as opposed to individual) sensory inputs (sight, sound, taste, touch, and smell) derived from our interactions with our spatial environment (e.g., Kirsh, 1995; Cooper & Lang, 1996; Barnes & McNaughton, 1985). The finding that information acquired via multiple senses lasts longer in memory and is more easily recalled than memories acquired by only a single sense is one of the foundational insights of multimodal learning theory (see Kress et al., 2001).

social· collective experience

A second set of approaches to studying memory that offers important insights for technical communicators is one that builds on the social view of memory.[3] In the social view, the memories of individuals and groups are inextricably intertwined and shared. As Fentress and Wickham (1992, p. x) put it, "One's private memories, and even the cognitive process of remembering, contain much that is social in origin." In this view, individuals and groups store up shared memories through collective experience and then recall these memories together by talking and writing about them. Memory work is best studied by examining "shared meanings and remembered images" of the collective: the family, the organization, the society (Fentress & Wickham, 1992, p. 59).

Approaches to studying social memory in the kinds of organizational contexts that concern us here go by different labels, including "knowledge management," "situated learning," and "organizational learning." The distinguishing characteristic of these types of approaches is that they generally do not study memory processes in laboratories or under controlled conditions but rather in the real-world contexts in which learning and activities occur, such as in schools and workplaces.

A central concern of knowledge management research is to understand how raw data or information becomes, or is transformed into, knowledge that is useful for individuals and groups. The diverse names of the theoretical focuses that inform this type of research signal the source of the knowledge-creating processes that each focus identifies as most vital. Organizational learning research often focuses on the phenomena of tacit knowledge (that is, unspoken, unwritten knowledge of how to do something held

Hidden unspoken

tacit → explicit

in the memories of an individual) and the various ways in which this knowledge can be extracted from or by the individual and rendered explicit, making it available to the rest of an organization (e.g., Zuboff, 1988; Nonaka &
Konno, 1998; Lam, 2000; DeLong, 2004; Kikoski & Kikoski, 2004). Any technical communicator who has interviewed a subject matter expert in order
to document a process or procedure has had the experience of rendering
tacit knowledge explicit, so these concepts can be important for technical
communicators interested in balancing memory with information overload.

One especially critical social approach to memory from which I draw
is situated-learning theory. The central tenet of situated-learning theory
is that human cognition is fundamentally place and circumstance based;
that is, human knowledge or, more accurately, our ability to understand
and interpret the world around us and our capacity to retain knowledge
in memory, is tied to the situations and circumstances in which we learn
this knowledge. Knowledge, then, "indexes the situation in which it arises
and is used": we retain information and recall it by associating it with the
physical, digital, and social circumstances that we perceive surrounding
it (Brown, Collins, & Duguid, 1989, p. 36). Because learning situations invariably include social affordances—shared workspaces, collective understandings, and shared texts—situated-learning theory informs a great deal
of research on communities of practice. Communities of practice are social
groups in organizations focused on shared work tasks. Situated-learning
theory argues that newcomers to communities of practice gradually become full members by, in a sense, acquiring a set of collective memories
and stories from their shared experiences working beside their more senior
colleagues (e.g., Lave, 1991; Lave & Wenger, 1991; Davenport & Hall, 2002;
Deuten & Rip, 2000; Brown & Duguid, 2002). Communal workspaces then
become scenes of distributed cognition, places in which groups of people coordinate their thinking during activities by utilizing language and other resources that they perceive in their shared physical contexts (e.g., Hutchins,
1995a, 1995b). Perhaps because so much of technical communicators' work
occurs in complex organizational communities of practice, situated approaches have proven especially helpful for understanding issues of information, knowledge, and memory and have been employed in a variety of
studies of learning to write in workplaces contexts (e.g., Blakeslee, 1997;
Winsor, 1996, 2001, 2003). Indeed, because the present study focuses on
precisely these sorts of workplace contexts, I borrow heavily from theories
of situated learning and distributed cognition and will be discussing these
theories in some detail at various points.

A third set of approaches to information, knowledge, and memory that contains important insights for technical communicators might usefully be labeled the "technological perspective" (Wick, 2000, p. 517). Where cognitive and social approaches are principally descriptive, the technological perspective is more concerned with design—specifically, the evaluation of external memory systems: the software and hardware interfaces, workspaces, and communication tools through which information offloading occurs in contemporary society and its workplaces. For instance, organizational memory research investigates the various high- and low-tech repositories in which an organization's information about itself is stored (e.g., Bannon & Kuutti, 1996; Walsh & Ungson, 1991). This research is primarily concerned with investigating how long-term repositories of information can support individual (e.g., Ackerman & Halverson, 1998) and group (e.g., Perry, Fruchter, & Rosenberg, 1999) work activities. One of the key insights from this research has been that, in most organizations, there is no such thing as a monolithic organizational memory. Rather, an organization's knowledge resides in the many smaller storage spaces provided by the digital, material, and human infrastructure (e.g., Ackerman & Halverson, 2000). The challenge for both the organization and the individual then becomes, as any technical communicator knows very well, one of finding the right repository containing the right knowledge to answer the right question at the right time.

Similar to studies of organizational memory, another technological perspective on memory comes from the fields of library and archival science, which tend to study the challenges of storing and accessing information over the long term, sometimes the *very* long term. Because archivists and librarians have been particularly affected by the evolution of the Internet and its ability to make carefully curated collections available with the click of a mouse, the research done in these fields is useful for studying the political and social issues entailed in accessing information (e.g., Delmas, 2001; Besser, 2002; Hedstrom, 2002; Bizjak, 2000). A central insight of this research is that merely making information available in databases does not necessarily make it usable to the individuals or communities who need it (e.g., Larsen & Wactlar, 2003). Such insights, of course, are quite familiar to technical communicators, who have experience with documentation processes driven by content management systems (e.g., Whittemore, 2008).

The various journals, special interest groups, and conferences of the Association for Computing Machinery (ACM) provide research that is relevant to discussions of information from a technical communication perspective. Generally, these approaches focus more on interfaces for automating everyday office information tasks than on storing information over the long term. For instance, Malone's (1983) study "How Do People Organize Their Desks? Implications for the Design of Office Information Systems" introduced a framework for studying workspace information in terms of finding, reminding, browsing, and searching that remains relevant. While some of this research is highly theoretical, seeking, like Rice, McCreadie, & Chang (2001), to describe searching and browsing as fundamental human activities, much of it consists of empirical work that attempts to observe humans' interactions with their work- and information spaces with the goal of creating better metaphors than the desktop and better methods of organizing information digitally than via files and folders (e.g., Barreau & Nardi, 1995; Fertig, Freeman, & Gelernter, 1996; Thayer & Steenkiste, 2003; Zhang & Marchionini, 2005; Krishnan & Jones, 2005; Fussell, Kraut, & Siegel, 2000; Spinelli, Perry, & O'Hara, 2005; Freeman & Gelernter, 2007; Ravasio & Tscherter, 2007). However, despite the diversity and ingenuity of these efforts, none of the alternatives have yet gained widespread popularity, and the desktop and file folder metaphors continue to hold sway in computer interface design.[4]

A fourth approach to studying issues of information and memory goes under the name "personal information management" or PIM. PIM encompasses "both the practice and the study of the activities people perform to acquire, organize, maintain, retrieve, use, and control the distribution of information items" (Jones & Teevan, 2007, p. 3). PIM attempts to find common ground among the various strands of memory research in order to formulate a shared framework for researching and designing technologies to support the varied activities and processes entailed in information acquisition, memory construction, and knowledge formation and dissemination. In doing so, PIM limits its objects of study to "information items," or information that has been "packaged" into physical or digital form: "A hallway conversation . . . conveys information but is not itself a packaging of information. A conversation is not an information item" (Jones & Teevan, 2007, p. 8). To explore the circulation of these information items, PIM defines a taxonomy of the fundamental activities by which people manage information items in the following activities: finding, refinding, keeping, and meta-level activities of organizing—a useful high-level way of categorizing the activities we associate with working with information. In fact, this taxonomy is perhaps closest to those activities we associate with the term "managing information," and the approach of this book builds on the activities it identifies. However, it seems to me to be a mistake to too abruptly limit the study of information only to "packaged" information, because the evolving work of technical communication necessitates that technical communicators pay

attention to information in all its varieties: by listening to stories, by participating in conversations, and by engaging in the full social life of their teams.
In this regard, I disagree with Jones and Teevan: A conversation may not
itself necessarily *be* an information item, but it certainly can contain information items, particularly in the case of, for example, two or more technical
communicators comparing notes or exchanging points of individual information or knowledge with respect to solving a specific problem.

This brief summary of approaches to information and memory, of
course, barely begins to scratch the surface of the vast literature on the
topic. I submit, however, that the underlying premise behind each approach
is that human beings principally make sense of information by *offloading*
it in some way to their surrounding physical and social environments—
through writing, through talking, through manipulating tools. We offload
memory to our material environment through our uses of our tools and
workspaces; we offload to our social environment by sharing and communicating information and knowledge with various audiences, from peers to
consumer-customers. In other words, we manage information by making
it memorable, and this memory work does not take place solely or even
mostly inside our heads.

Together, however, the very multiplicity of these approaches hints that
we are not confident that we fully understand our subject. No single field,
in other words, "owns" these issues, and technical communicators therefore have an obligation to contribute to the discussion. In short, despite
the considerable insights offered by this vast and ever-expanding literature
on cognition, information, knowledge, memory, and IT interfaces, none of
these approaches speaks directly to the innovative information-managing
work that the evolving career of technical communication entails. This innovative work requires that technical communicators manage information
to fuel not only their own creative processes but also those of their teams
or entire organizations, each with its own set of cultural assumptions about
information and knowledge.

This information lives in many places: in our own and our colleagues'
organic memories (cognitive perspectives); in documents, texts, and other
"information items" (information management perspectives); in stories or
narratives told among teammates (narrative and communities-of-practice
perspectives); in material infrastructures (situated and joint cognition perspectives); and in our physical and digital tools and databases (technology
perspectives). In the face of such complex challenges, technical communicators need strategies for inquiring into, understanding, and working productively within the unique information cultures of their organizations.

This strategic framework should be based on theories outlining what constitutes useful information for technical communicators and what role writing, still perhaps a technical communicator's most distinctive skill, plays in processes of transforming information into knowledge. Without such theory- and research-based strategies for inquiry, technical communicators are forced to rely on an idiosyncratic set of tips, tricks, and best practices for managing information that they have picked up in school, during their careers, or from research that speaks only indirectly to their concerns.

INFORMATION MANAGEMENT ACTIVITIES ARE RHETORICAL MEMORY PRACTICES

I propose that an approach to theorizing and studying information, knowledge, and memory more appropriate to the current needs of professional and technical communicators can be found in rhetorical theory: we must begin by reconceptualizing information management practices as *rhetorical memory practices*. The three words that define this term help define our focus. First, this approach is *rhetorical* because rhetorical theory has always asserted that memory work—the manipulation of stored information—plays a central role in creative processes of communication. Like technical communication, rhetoric is a productive art, an art concerned with making or creating new meanings through words and other forms of communication. In rhetoric's original formulations in ancient Greece and Rome, memory was given pride of place as one of the five interconnected divisions, or "canons," of rhetoric that describe the creative process. As one of these essential divisions, memory was honored as "not an alternative to creativity . . . but the route to it" (Carruthers, 1990, p. 192).

Rhetorical theory holds that the process of retrieving and adapting existing knowledge to the exigencies of shifting communication situations is absolutely essential to the creative process by which communicators determine what to say and how to say it to meet the needs of their audiences. A rhetorical view of memory accords with cognitive psychology in refuting both the popular notion that true creativity arises not from old information but from wholly original bursts of inspiration (Carruthers, 1990) and the pervasive idea that the operations of human memory are somehow analogous to using a computer database, a matter of simple regurgitation (G. Johnson, 1991).

Recent neurobiological research on cognition adds weight to the rhetorical view of memory through the discovery that even the most obdurate long-term memories are, in fact, dynamic and malleable and can translate or transform past experience into new forms based on situations of remembering (e.g., Nader, 2003). A rhetorical theory of memory, therefore, recognizes the creative value to an organization of precisely those kinds of information-manipulating work that technical communicators are now finding themselves called upon to do: finding, assessing, filtering, and translating information to make useful, actionable, memorable knowledge for diverse audiences in contingent and variable situations. Memory work, in a rhetorical sense, is more than being a thorough organizer or recalling details with the accuracy of a human iPod; it is about enabling new thoughts and ideas to arise from stored material, as well as to be recombined with new information in the creation of new knowledge and ideas. In other words, memory work in this rhetorical sense is about drawing information from the entire broad collective of information systems at our disposal, both mental and physical.

Second, *memory practices* describe phenomena that are elsewhere variously termed "information managing," "knowledge work," or "symbolic analysis." The term "practice" highlights the social and embodied aspects of memory work, aspects that can be overlooked if we focus too narrowly on this work as manipulating things (i.e., information items) rather than as part of the process of participating in social and workplace life. Social theories of activity define a practice as "a set of socially defined ways of doing things in a specific domain" (Wenger, McDermott, & Snyder, 2002, p. 38). In other words, our methods for making information memorable—for doing memory work—are socially conditioned approaches to common problems presented by information. Even our most apparently original and idiosyncratic techniques for storing or retrieving information are rooted in and conditioned by the larger cultures in which we live and, in the case of technical communication, in the organizational cultures of our employers, and further, to some extent at least, in the culture of our workplace predecessors and coworkers.

The term "practice" also points to the effects of this social conditioning on our bodies when we do memory work. Practices, according to Bourdieu (1990, p. 57), "exploit the body's readiness to take seriously the performative magic of the social." As the knowledge management research makes clear, our practices for storing and retrieving information and making knowledge rarely occur solely in our heads—they almost always have a physical, embodied component. Even when we seek information in our organic memories, we often do things with our bodies like hitting our heads or making gestures with our hands. If such gestures are somewhat idiosyncratic and not connected in a logical way to the actual process of retrieval, consider the ways in which our manipulation of external information often entails using our bodies to interact with our tools and workspace environments. To borrow (and slightly repurpose) an example from Hutchins's (1995b) research on distributed cognition, consider an airline pilot operating a 747 jetliner. To an outside observer, the cockpit presents an overwhelming array of dials and gauges, toggle switches, levers, and controls. But the veteran pilot competently and very confidently operates the plane through a series of fluid motions, eyes, hands, and feet flowing over the controls with the precision of ballet, such that the performance of flying the plane in knowing which switch to hit next is almost as much about the embodied, physical aspect of memory as it is about the explicit knowledge of what that switch actually does. And while it is true that our typical office workstations are not nearly so complicated as the cockpit of a jetliner, nevertheless, there are physical motions in the operation of our computers and other equipment and communication devices with which we become almost unconsciously familiarized in the same way as the airline pilot, and through which we orchestrate our workplace performance. Each of these gestures, postures, movements, and manipulations is afforded or constrained by the tools, spaces, people, and infrastructures in which we find ourselves embedded—in other words, by our memory culture.

Most importantly, though, the notion of memory practices as manifestations of a memory culture introduces one of the foundational concepts of this book: the memory regime. According to Bowker (2005, p. 9), who coins the term, memory regimes "articulate technologies and practices into relatively historically constant sets of memory practices that permit both the creation of a continuous, useful past and the transmission sub rosa of information, stories, and practices from our wild, discontinuous, ever-changing past." In addition to its more obvious manifestation in the set of computer interfaces, database technologies, and built infrastructures that an organization selects for managing it memories, an organization's memory regime shapes what types of information have value for creating knowledge and what types of information can be used as warrants for arguments. In other words, a memory regime has politics that affect things like what is preserved in and what is deleted from the collective memory, who has access to the information held in the collective memory and who does not, and who is empowered to participate in the collective memory by telling the organization's stories and who is not. The local politics of memory is a confounding aspect of managing information that is, perhaps as a consequence, too often elided or ignored, but it is one that technical communicators must be prepared to understand.

tools + practices

Throughout the book, I will use the term "memory regime" to refer to the collective set of memory practices, the memory culture, as it were, of an organization, and I illustrate the concept with specific examples from the organizational case study that serves as the foundation for this book. In examining these ideas, I attempt to understand the memory regime at one software company—Software Unlimited—by pursuing answers to the following questions:

> What types of organizational information are the most important for technical communicators?
>
> Where does this information reside and how does it move through the organization?
>
> How do technical communicators transform information into useful knowledge?
>
> How does this memory work contribute to the status and professional identity of technical communicators within their organizations?
>
> Finally, why—why does memory work matter? Why, if at all, does it enhance the status and prestige of technical communicators in their workplaces?

Additionally, although there are limits to what a single case study can offer, I conceive the value of this volume to lie in helping academic audiences see and understand

> the usefulness of conceptualizing information managing work as social practices and, specifically, as rhetorical memory practices;
>
> the value of the methodological framework I articulate in chapters 2 and 3 to helping study and interpret memory work in organizations;
>
> what an art of memory looks like in the hands of skilled but relatively typical practitioners at various points along their career paths and with their communities of practice; and
>
> issues of power, status, identity, and agency arising from collective and emergent memory practices.

CHAPTER TWO

Memory Work as Embodied Rhetorical Practice

show an approach to studying rhetorical invention.

> Knowledge is something we digest rather than merely hold. It entails the
> knower's understanding and some degree of commitment.
> —John Seely Brown & Paul Duguid, 2002, p. 120

The goals of this chapter are to articulate a working description of a rhetorical memory practice and to ground this description in rhetorical
theory and contemporary psychology in order to arrive at an approach for
studying such practices as components of rhetorical invention. As I noted
in the previous chapter, Bowker (2005) employs the term "memory regime"
as a way of describing the *collective* (or the complete collection of) memory
practices of a given culture. The function and effects of a memory regime,
therefore, can be equated with the function and effects of culture. Specifically, similar to what Slack and Wise (2005, p. 4.) note about culture, a
memory regime performs "the work of selection: the selecting, challenging,
arranging, and living of [the] received artifacts of everyday life." That is,
like culture, the memory regime determines what can and cannot be said,
what counts as knowledge, and what evidence can be used as warrants in
arguments and debates. As cultures have subcultures, which both partake
of the features of the larger culture and adapt those features for local circumstances, memory regimes also have specific instantiations, such as in
organizations that both adopt and adapt the memory tools and practices of
the larger culture to meet specific organizational needs.

As we will see in this study, organizations are also prone to "contain"
multiple sub-memory-regimes or nested memory regimes within their overall corporate structure. An easy example of this phenomenon might be a retail department store like Macy's that has multiple locations spread across
an entire continent. It is reasonable to expect that the need to cater to dif-

ferent ethnic and demographic populations would entail some differences in the memory regime of each region and even of each location. In fact, the average income of the local population alone is sufficient to warrant significant contrasting differences in the way specific stores market their products, not to mention the specific lines of products each location might carry in order to cater to its specific customer base. Further, with its nearly eight hundred locations throughout the United States, Macy's has some locations that are union shops, while the majority are nonunion. There can be little doubt that the memory regimes will differ based on significant determining factors—like whether a particular location is union or nonunion—that are built into the very structure of an organization.

With that in mind, however, and in accord with research studying the development and propagation of organizational subcultures (e.g., Winsor, 2003; Hallett, 2003), the present study indicates that such sub-memory-regimes also exist in much smaller companies, with varying degrees of success. This is especially true for companies that experience explosive growth, wherein there is little or no time for the establishment of strict bureaucratic protocols (which some entrepreneurs would probably regard as a very good thing), or for companies whose main business focuses on exploiting rapidly developing cutting-edge innovation in an assortment of areas such as marketing strategies (consider Amazon or Google or Zappo's when they first started out) or on emerging—and highly unpredictable—high technology, wherein any effort to impose a strict, top-to-bottom corporate memory regime might be regarded as constricting or stifling creative, free-form innovation, invention, and experimentation. Such, indeed, appears to be the case at Software Unlimited, which, for reasons that will be explored throughout the book, eschews a strict, overarching memory regime. However, for the purpose in this book of examining and understanding the rhetorical memory practices of information managers and technical communicators, before exploring the concept of the memory regime, we must first drill down to the level of the individual who performs memory work under or within the framework of the company's memory regime (or regimes) in the performance of his or her job.

Again, like culture, the memory regime itself is most readily discernible at the level of practice, which "denotes a set of socially defined ways of doing things in a specific domain: a set of common approaches and shared standards that create a basis for action, communication, problem solving, performance, and accountability" (Wenger, McDermott, & Snyder, 2002, p. 39). Further, *individual* memory practices are the activities and tools by which members of a given memory regime attempt to deal with stored information from the past and an ever-increasing influx of new information from multiple sources, including a large variety of activities ranging from note taking to data basing. These activities and tools are substantially influenced by the memory regimes in which they reside, but they are also often employed in idiosyncratic ways in actual work processes based on the experience and the intelligence—both cunning and practical—of individuals as they respond to situations. For example, as the present research study reveals, some participants handwrite notes in notebooks during meetings so that they can later refer back to those notes while composing, while others do so primarily as a method of imprinting the material more firmly into their own long-term memories—findings similar to those of Ann Blair (2004) in her study of note-taking practices throughout history.

Whether or not it is practical (or indeed, even possible) to study individual memory practices independently of the particular memory regimes under which information managers (including, of course, technical communicators) operate, it is important to study them within the context of the memory regimes in question (i.e., the memory regime of Software Unlimited and the subregimes of its work teams) for a number of reasons. To begin with, workers—at least those ambitious workers who desire to perform their jobs well as a means of helping their company grow and prosper—devise and employ their own unique sets of memory practices in order to succeed in (and be recognized for) their job performance, thus helping the company to succeed. In order to achieve success in their job performance, typically workers must create, adopt, or adapt reliable memory practices that support not only the goals of the organization, but also its standards of operation, all of which, in turn, are embodied in the memory regime. Thus, the memory practices of individual workers are both guided by, and enacted in direct support of, the prevailing memory regime (or regimes) of the organization. For this reason, it would be impossible to fully understand individual memory practices without appeal to the whys and wherefores for which they are enacted by individual workers. In one sense, albeit to a limited degree, they are opposite sides of the same coin, although often allowing for a large measure of individual preference, creativity, choice, and adaptation of memory practices among individual workers—adaptations permitted as long as those practices appear to further the company's prosperity and goals. In another sense, we may be able to say that individual memory practices operate as functions or instantiations of the memory regime.

Further, while we should be careful not to conflate individual memory practices with the memory regimes in which they reside, the two con-

cepts are not separable in our analysis—it would be impossible to describe a
memory regime without first describing the often diverse array of practices
occurring under its aegis, practices which both describe and constitute the
regime. In other words, regimes and practices are coarticulatory, a relationships that I will discuss further in chapter 3. At this point, it will suffice to
note that memory regimes and the individual memory practices through
which they are discernible are connected via the concept of tool mediation
as described by the Soviet psychologist Lev Vygotsky (1978): "Tools transform natural mental processes into instrumental acts . . . that is, mental
processes mediated by culturally developed means" (Kaptelinin & Nardi,
2006, p. 42). Tool mediation, in short, describes how the outside gets inside,
or, as Engeström (1999, p. 29) glosses Vygotsky, "Mediation by tools and
signs is not merely a psychological idea. It is an idea that breaks down the
Cartesian walls that isolate the individual mind from the culture and the
society."

Thus, this study endeavors to identify the memory practices of information managers as they work—the articulations by which these *symbolic
analysts* (Reich, 1991) overcome information overload via creative and
sometimes idiosyncratic use of mediational assemblages of both high- and
low-technology tools offered by their current work contexts but also influenced by psychological tools and internal "devices" derived from previous
experience. This study further comprises an attempt to closely observe and
understand the complex approaches to memory work through which information managers and technical communicators in contemporary workplaces create knowledge by retrieving and adapting stored information and
assimilating new information to meet the exigencies of rhetorical situations. In other words, it conceives of memory work as rhetorical practice
and investigates the role in composing and writing of "high" technologies
like computer interfaces and databases *in tandem* with "low" technological mediations offered by human embodied interactions in space and time
for the purposes of changing memory regimes and informing the design of
more effective technologies in the future. So we begin at the level of the
individual information manager and his or her preferred choice and utiliza-
tion of memory practices within specific embodied contexts.

KNOWLEDGE IN THE RHETORICAL MEMORY TRADITION

To gain an understanding of the modern memory practices of the information managers at Software Unlimited, who are also technical communicators, and further, in order to analyze how current-day memory techniques

or practices are rhetorical in nature, we should understand how the founders of the rhetorical tradition first theorized the role of memory work in the art of rhetoric. For this purpose we principally turn to Aristotle. But, before grappling with such a titanic figure, let's take brief a step back by beginning with a basic definition of "practice." *The Oxford English Dictionary* (1989) provides a definition of *practice* as being (among others) "the action of doing something; performance, execution; working, etc." The importance of this modern definition of practice in the context of our discussion is that a "practice" entails some sort of action being taken. On the other hand, we tend to think of memory—insofar as it involves information or knowledge—as static, particularly in our modern work environment of electronic data files (whether large documents or snippets like emails and tweets), wherein knowledge or information seems somehow housed in discrete "packets" to be filed or stored either electronically or physically in an office file cabinet. However, what the very definition of the word "practice" tells us is that a memory practice involves both the information or knowledge itself and some action taken with or based on that information or knowledge, by individuals: with respect to the present study, by the information managers at Software Unlimited.

Saugstad (2002) posits the claim that Aristotle had both "a broader understanding of knowledge than we usually have today" (p. 377) and a more "sophisticated description of practice" (p. 378), also, presumably, than we typically have today. Let's examine both claims. First, Saugstad states that our "contemporary notion of knowledge is influenced by the medium of writing, where knowledge . . . can be conceptualised and formulated in writing" (p. 377). The Aristotelian understanding of knowledge, Saugstad goes on to say, "captures many of the intuitive, bodily, and experience-based forms of knowledge" (p. 377). Saugstad may have a point in that we tend to regard knowledge, or the "stuff" of knowledge, as a kind of one-dimensional commodity to be saved and stored. One thinks of the numerous encyclopedias published over the past two centuries under the rather audacious rubric of "The Book of Knowledge," as if all knowledge could be squeezed into the pages of a single or even a multivolume book.

But Saugstad's claim that our contemporary notion of knowledge is influenced by writing, or the medium of writing to be exact, is both provocative and potentially very insightful for the present discussion of rhetorical memory practices as exhibited by the information managers in a modern-day company, who are the subjects of this study. First, as I note later in this chapter, rhetorical memory practices were originally the purview of orators as a mnemonic means of remembering both the content and the dramatic

presentation of speeches, and rhetoricians made use of visual cues in the
environment as well as their own bodies as mnemonic devices. With that
in mind, it is obvious that writing represents the first-ever infrastructural
memory practice, involving, as it does, the action of the actor, cognition and
information or knowledge, and the necessary (infrastructural) out-of-body
physical objects—a stylus of some sort, and something to write on, be it a
cave wall or a sheet of papyrus—or a computer keyboard. Everything that
has come after writing with a pen and paper is but a variant of the basic and
fundamental process of writing. Computers and other devices may indeed
be more technologically sophisticated, faster, more efficient, and so on, but
working a keyboard is still, at bottom, writing. This explains why it is important to talk about writing in the context of a study devoted to rhetorical
memory practices, and further, why it is important to understand writing as
an embodied practice precisely in the way in which we, and Aristotle, have
defined practice in the foregoing. Arguably, then, in an important sense
everything that one uses in service to a memory practice—and by "everything" in this context, I mean all of the tools, devices, and affordances of
the modern office environment—is an extension of the mind and body, and
hence, the very definition of an embodied activity or practice.

tools

   More to the point, however, Saugstad (2002, p. 376) prefaces her remarks
about Aristotle's broader concept of knowledge by stating that "the Aristotelian categories of knowledge . . . are difficult to fit into the modern world
dominated by advanced technology" and that "technology has changed
practical life." While Saugstad seems to be suggesting that technology may
have caused some of us to narrow our conception of "knowledge" to, in
essence, extend or align more closely with something resembling "data,"
perhaps without consciously realizing it, it does not take a great leap of
logical analysis to realize that knowledge is much more sophisticated, and
in fact even in contemporary times and despite modern technology, our
concept of knowledge today must still entail the Aristotelian components
of intuitive, bodily, and experience-based forms. Moreover, in this age of
the explosive proliferation of information communicated instantaneously
across the world, information managers of all stripes are learning just how
broad and how sophisticated the concept of knowledge actually is. And this
is especially true for technical communicators in their effort to stay on top
of this information overload and to make sense of that information so as to
convert it into genuine knowledge.

   Saugstad (2002) argues that Aristotle's description of practice, like his
broader conception of knowledge, is also more sophisticated than the dictionary definition given earlier, or our common usage today: "Aristotle

shows that practice appears less predictable and more knowledge-rich, varied and perspective-filled than convention today perceives it" (p. 378). And indeed, the role of these other dynamic, less certain, and sometimes conflicting aspects of practice will be evident in my descriptions of the memory work and the interactions among the case study participants and their coworkers at Software Unlimited. Finally, in a manner of bridging the connection between practice and knowledge, Saugstad asserts "Aristotle's understanding of knowledge as human activity. . . . Aristotle . . . perceives knowledge as a competence, as something you are or something you do" (p. 378). As we will see in the case study chapters, this very "human activity," in which the information and knowledge possessed by the information managers and their coworkers varies widely, and where the competences of the players involved in performing their jobs also varies considerably from one individual to another, is also evident in their interaction with each other and as a group. In fact, as I will discuss in more detail below, Aristotle's definition presages contemporary social theory's conceptualization of practice as "a set of common approaches and shared standards that create a basis for action, communication, problem solving, performance, and accountability" (Wenger, McDermott, & Snyder, 2002, p. 39).

And as we focus our attention on the participants in this case study, two other Aristotelian concepts will become very important in our discussion; concepts that speak to this level of competence, specifically, the kind of Aristotelian notion of competence driven by knowledge, experience, and individual action or practice. These two concepts are the Aristotelian notions of *kairos* and *hexis*.

As I noted earlier in this chapter, and will explain in greater detail in chapter 3, I assert that Software Unlimited at the time of this study is an example of a paradigmatic, fast-paced, high-technology company whose very existence depends on being, always, at the forefront of innovation both in terms of creating new software packages for its customers, while also regularly adding new features to its current stable of software product offerings to keep them "new"—and therefore desirable—to their customer base. Like any other software company, the working environment at Software Unlimited is very much a high-pressure, think-on-the-fly (and do-on-the-fly) atmosphere—and, often, a very exciting one—full of novel, perhaps unorthodox ideas and unique concepts dreamed up by a highly creative staff. It is also an *internally competitive* atmosphere, and one need only think of Apple's determination to introduce a new "generation" of the iPhone every year—and the media buzz that attends each new iteration—to understand what that environment must be like.

As I have also alluded earlier in this chapter, Software Unlimited is guided and driven by a subset of largely independent but overlapping memory regimes that roughly correspond to a similar subset of communities of practice—a subset of memory regimes that, like the creative individuals working for the company, are, to a significant degree, in a state of competition with each other. It is within this internally competitive environment that the information developers observed in this study must strive to learn the overall company ethos, of course, but most importantly in the context of this study, to develop their own strategies and tactics to elevate their individual levels of knowledge, expertise, and practice, thereby elevating their stature and influence among coworkers (both within and outside their specific communities of practice). Such achievement enhances their power and ascendancy within the company hierarchy.[1] These are, after all, ambitious individuals, and while matters of the corporate hierarchy are beyond the scope of this book, this free-form, competitive situation affords a good opportunity for studying the unique memory practices employed by information-managing technical communicators, with varying levels of experience and job tenure, in the performance of their job duties. Stated more bluntly, if these information developers want to win influence and gain stature among their colleagues—as well as have their own ideas and innovations rule the day, they've got to bring their best game to every eventuality, from digital documentary sources to group meetings where decisions are made. To understand how these information developers endeavor to accomplish this through the development of individual memory practices, and, more importantly for our purposes, how the processes they employ in this effort are rhetorical in nature, we must first examine two of Aristotle's categories of knowledge, *techne* and *pronesis*, and their relationship to individual competence within rhetorical situations.

FOUR KEY CONCEPTS FROM RHETORICAL MEMORY THEORY: TECHNE, KAIROS, PHRONESIS, AND HEXIS

The distinctions that Aristotle makes between techne knowledge and phronesis knowledge can help us to understand the rhetorical work that memory practices enable us to accomplish. Aristotle distinguishes theoretical or epistemic knowledge—knowledge concerned with general or universal principles—from practical knowledge, knowledge of how to act or create within specific situations in the world of lived experience. Because our focus is practice, it is primarily the latter type of knowledge that concerns us here. Aristotle further specifies two types of practical knowledge roughly distinguishable based on their respective *teloi,* or ends. The end of techne knowledge is "the thing made," the artifact or output of the application of the knowledge (R. R. Johnson, 2010, p. 678). Thus, techne knowledge encompasses "the kind of knowledge possessed by an expert in [a] specialized craft [including] the principles underlying the production of an object or state of affairs, e.g., a house, a table, a safe journey, or a state of being healthy" (Dunne, 1993, p. 244). In the context of this study, *techne* refers to knowledge of the technical details of the job and its processes—relating to memory practices that contribute to learning to do, or learning how to do, by enumerating all of the potential components, without necessarily knowing how all of those pieces fit together. As Saugstad (2002, p. 380) puts it,

> *like*
> *how to*
> *complete a*
> *journal*
> *entry in*
> *accounting*
>
> Techne is more than a competence, as it both consists of an ability to carry out a procedure in practice . . . and to give an account of the general principles behind the procedure. A person with techne can learn the general principles behind the procedure and in this sense techne is a reflective knowledge about general aspects. However, unless one has training and experience one is not able to apply the general principles to the given specific situation.

For the purposes of this study, however, it is instructive to consider techne knowledge in conjunction with another concept that was crucial to the ancient rhetoricians—that of kairos. Kinneavy (1986) interprets kairos as a two-part concept in Roman thought emphasizing both time and appropriateness: "the right or opportune time to do something, or right measure in doing something" (p. 80). *the right time* Time and appropriateness together, Kinneavy points out, delineate the boundaries of a rhetorical situation: "the appropriateness of the discourse to the particular circumstances of the time, place, speaker, and audience involved" (p. 84). For the master of a techne, "success is . . . achieved . . . by a flexible kind of responsiveness to the dynamism of the material itself. . . . This is the meaning of grasping the kairos; one's active intervention has skillfully waited until one's polyvalent materials . . . are at their most propitious" (Dunne, 1993, p. 256). Mastering a techne, then, means making oneself as far as possible "invulnerable to chance" within one's task domain (Dunne, 1993, p. 256). The end of a techne lies in the thing produced, but the measure of one's mastery lies in responsiveness to kairos. *appropriate time to do something?*

In practical terms, when we consider the memory work that must be accomplished by the information developers at Software Unlimited in conjunction with these two important rhetorical concepts—to develop techne

by learning from the substantial volume of technical information or knowledge that they must bring to bear, and kairos, the necessity to access and
present the right information at the right time in the performance of their
work—we gain a strong affirmation of Aristotle's broader understanding
of knowledge as "human activity," and as Saugstad (2002), quoted earlier,
further describes it, "knowledge as a competence, as something you are or
something you do" (p. 378). Saugstad concludes that techne "is primarily
learnt by doing that which one is to learn" (p. 380), and in its classical origins, the rhetorical canon of memory was in fact primarily concerned with
the larger practice of *retrieving* stored information specifically for *kairotic*
purposes in order to facilitate invention during rhetorical situations. As
Carruthers (1990, p. 19) puts it, in classical and medieval rhetorical theory,
"the proof of good memory lies not in the simple retention of even in large
amounts of material; rather, it is the ability to move it about instantly, directly, and securely that is admired." Similarly, Crowley and Hawhee (2012,
pp. 331–32) emphasize the link between memory and kairos in processes of
rhetorical invention: "Memory was not only a system of recollection for
ancient and medieval peoples; it was a means of invention. In ancient and
medieval times, people with trained memories could memorize huge vol-
umes of information, along with keys to its organization, and carry all this
in their heads. Whenever the need arose to speak or write, they simply retrieved any relevant topics or commentary from their ordered places within
memory, reorganized and expanded on these, and added their own interpretations of the traditional material. People who had trained their memories
could do this sort of composing without using writing at all."

The second of Aristotle's categories of knowledge that we need to consider within the context of rhetorical memory practices is phronesis. Where
the ends of a techne lie in the products or artifacts created by the application of one's technical (techne) knowledge, "phronesis is the knowledge of
human action where the end (telos) is good action itself" (R. R. Johnson,
2010, p. 678). More simply stated and for our present purposes, we may
think of phronesis as encompassing techne but adding social, ethos-related
knowledge. In essence, phronesis encompasses a degree of mastery that
entails not simply a knowledge of the technical details of a practice, but
also a deeper, fundamental understanding of the social context and situated
ethical reasoning underlying the practice. In other words, where a person
possessing a techne may reach the right conclusion at the right time in
the kairotic moment, "the phronetic person reaches the right conclusion
at the right time *and on the basis of the right arguments*" (Saugstad, 2002,
p. 381, emphasis mine). Phronetic knowledge, in other words, is knowledge

that contributes to the acquisition of a good professional identity within one's organization or communities of practice. Further, a worker gains or acquires phronesis knowledge more as a matter of habituation than as matter of mere acquisition. That is, achieving phronesis knowledge entails both physical and mental habituation not implicit in techne: you cannot find phronetic knowledge in a database; you must inculcate it within yourself.

It is important for the discussion of embodiment and memory work below and for the case study that rhetorical theory describes the physical and mental habituation required for phronesis via the fourth concept I want to call attention to, the concept of hexis. *Hexis* is the term used by the ancient theorizers of rhetoric to refer to mastery in an art (i.e., a productive activity, a practice): "Quintilian defines hexis as that 'assured facility' (firma facilitas) in any art which supplements and transcends the rules themselves, and constitutes what we call mastery" (Carruthers, 1990, p. 69). *Hexis*, in other words, describes an essential (and, important for the researcher, frequently visible) component of phronetic knowledge: bodily disposition. And, critically, one's hexis as a necessary component of acquiring phronesis is achieved as a product of both mental and physical discipline (see fig. 2.1). As Hawhee (2004, p. 58) puts it, "For Aristotle . . . disposition is inexorably tied to thought—transformation of one inevitably produces transformation in the other. . . . Thought does not just happen within the body, it happens as the body." Further, as it is both mental and physical, hexis is always also both individual and social in nature. Bourdieu (1990) calls attention to this social aspect of hexis implicit in its ancient formulation: "Body hexis [is] linked to a whole system of techniques involving the body and tools, and charged with a host of social meanings and values" (p. 87). In short, Bourdieu concludes, "Bodily hexis is political mythology realized, em-bodied, turned into a permanent disposition, a durable manner of standing, speaking, and thereby of feeling and thinking" (pp. 93–94).

Phronesis and *hexis*, then, describe important aspects of social knowhow that, in the context of this study, enable an individual to achieve both stature among and influence over coworkers by virtue of both superior technical knowledge and a learned and practiced ability to negotiate effectively with those coworkers in decision- and policy-making situations. From this description, we can conclude that *phronetic knowledge* describes the kind of, often tacit, knowledge that one can gain only through extended and situated dwelling within a memory regime and community of practice, while *phronesis* represents the state that ambitious employees strive to achieve both for the benefit of their own career advancement and for the growth and prosperity of the companies they work for. Last, the concept of hexis

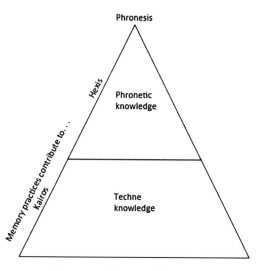

Figure 2.1 Elements of mastery.

demonstrates that the state of phronesis cannot be achieved through mental activity alone (i.e., through explicit learning—as in reading the company's orientation manuals or participating in its training course), but only via active, appropriate, ethical, and embodied participation in a community's activities.

Habits, however, can be changed or adapted to fit new or altered circumstances. Similarly, an effective bodily hexis, though a product of habit, can be leveraged to contribute to learning and the acquisition of new hexeis. An effective hexis entails a receptivity to change that can make learning under new or changing circumstances easier (see fig. 2.2). As Carruthers (1990, pp. 69–71) points out, while a hexis is "physiological, as the memory is trained to respond to certain movements, just as a dancer's muscles are," it can also be "reasoned, for it is a 'facilitated' rather than an 'automatic' response. . . . For the trained memory was not considered to be merely practical 'know-how,' a useful gimmick that one might indulge in or not. . . . It was co-extensive with wisdom and knowledge." Such a hexis would be necessary to maintaining one's phronesis over the long term in one's workplace, as not just new software and hardware, but new people, new ideas, and new ways of doing things were introduced.

In evaluating the effectiveness of information-managing activities as rhetorical memory practices, we might usefully, then, contrast those activities that principally contribute to techne knowledge with those that foster or otherwise contribute to phronesis knowledge. While information-

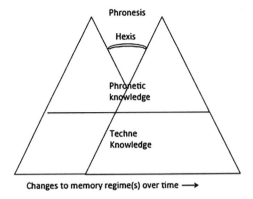

Phronesis

Hexis

Phronetic knowledge

Techne Knowledge

Changes to memory regime(s) over time ⟶

Figure 2.2 Hexis as bridge: An effective hexis leads to small discontinuities in phronesis.

managing work contributing to techne may help a person succeed in the short-term, kairotic moment, it does not necessarily contribute to his or her growth in phronesis over the longer term and could, under certain circumstances, even inhibit such growth. As just a brief example, consider the person who retains every email he or she ever receives in order to preserve a complete paper trail of responsibility for every decision (such activities are often acerbically described as "CYA," for "covering your ass"): such diligent archiving may certainly assist an individual in avoiding blame in a given situation, but it may lead to information overload, or, equally detrimental, its too-frequent use as an appeal in arguments may cause the individual to acquire a bad reputation as a shirker or nitpicker, the opposite of the phronetic person. Such a person may be a master technician (i.e., a master of the techne), but he or she will probably not have a good reputation among his or her communities of practice (in popular depictions, IT personnel are frequently caricatured in this role, such as the character "Mordac, The Preventer of Information Services," in Scott Adam's *Dilbert* comic strip). The phronetic information manager, by contrast, has learned which information to preserve and which to purge, and, more importantly, knows when, with whom, and, most critically, why to preserve or deploy a given piece of information.

Similarly, while an effective hexis—an embodied habit or disposition that proves beneficial to the individual or firm—must be a part in the growth of phronesis, not all embodied hexeis are good, nor, moreover, should even a good one be immune to change. In other words, hexeis must sometimes be unlearned. As Hawhee (2004, p. 58) puts it, "Hexis thus describes the bodily 'state' that enables particular kinds of cunning, intelligent

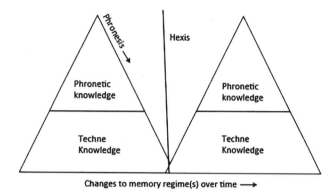

Figure 2.3 Hexis as barrier: An obdurate hexis can make adapting to change difficult, leading to loss of phronesis.

responses. What happens, though, if one cultivates multiple *hexeis*? . . . This capacity to change, to assume a new hexis . . . makes all the difference." Hawhee is pointing out that a capacity to learn and assume new hexeis may be the hallmark of a particularly beneficial kind of creativity, while the opposite—an obdurate clinging to an outmoded or superseded way of doing things—may become a stumbling block to creativity and, eventually, initiate a decline from a state of phronesis previously achieved (see fig. 2.3). Moreover, because they represent the "set of socially defined" ways of interfacing with past information in a given memory regime, memory practices may prove essential to both acquiring and, when necessary, changing hexeis: even something as simple as changing one's habitual method of retrieving information on one's personal computer from browsing to searching (or vice versa) may unlock heretofore-undiscovered insights or recover lost knowledge for the individual and, in potential at least, his or her work teams (Wenger, McDermott, & Snyder, 2002, p. 39).

DIMENSIONS OF A RHETORICAL PRACTICE OF MEMORY

Bowker (2005, p. 25) makes clear his view that the most important thing about memory practices is that they exist primarily to enable action in the present rather than to preserve a perfect record of the past: "The act of remembering . . . is one of our chief ways of dealing with the world as effective creatures: it is a way of framing the present; a mode of acting." Consequently, Bowker notes, memory practices are about forgetting as much as they are about remembering. Building on Derrida's (1998) theory of the archive, Bowker adds that memory practices tend to be both *sequential* and *jussive*. That is, memory practices are sequential because they partake in standardizing and classifying information so that it can be found when needed, and they are jussive because they often participate in the process by which information judged to be of no use is purged from memory—all in order to enable practical action in the present and future.

Further, according to Bowker, memory practices become incorporated into the built environments that surround us—our buildings, our work-spaces, and our digital spaces—in other words, into our infrastructures and tools. Alluding to his earlier work with Susan Leigh Star (Bowker & Star, 1999), Bowker (2005, p. 21) notes that standards and classifications, embodied in infrastructures, contain affordances by which we can offload part of the burden of memory: "We classify in order to be able to forget." Collectively, these affordances, when they are working properly, enable something rather like what I choose to refer to as a "standing wave" of memory whereby we are able to manage the information load necessary to accomplish our daily tasks without becoming overwhelmed by the work of recall—that is, without suffering information overload. This notion of a "standing wave" of memory echoes the "memory-ready" condition of classical rhetoric emphasized by Crowley and Hawhee (2012) and noted earlier. We are able to do this because, under ideal conditions, infrastructures offer affordances that allow us to offload memory so that precisely the right amount of information required to perform a given task presents itself at any one moment: "We are not in general able to remember complete stories about the past. . . . What we do well is to disaggregate a fact about the past into a number of standard elements, and then set in train a procedure for reassembling the specific out of the general. This sets in motion a system of memory recall that is able at any given moment to create a working version of the past" (Bowker, 2005, pp. 17–18).

In other words, for Bowker, successful memory work depends a great deal on situational affordances. Organizations, Bowker (2005, p. 15) notes, "delegate memory tasks to the environment." Further, these infrastructures aid memory by organizing and in many cases limiting what we can and cannot perceive or interact with in a given situation, phenomena related to cognitive *cuing* and *constraining*, as noted by psychologists studying user-centered design (Norman, 2002).

However, Bowker's (2005) methodology for studying memory is historical and textual rather than ethnographic or observational, so he focuses more on the larger systems of memory than on the actual practices by which individuals or groups "set in train" these procedures for recall during activities (p. 18). While it might, then, be said that Bowker's research gives a sense of the *where* of memory practices, Star's fieldwork methods enable her to get a detailed look at the *when* of memory practices in the real-time activities of her research participants. From her observations, Star (1999) formulates the concept of "articulation work" as a label for the "real-time adjustments" that people continually perform below the level of their visible work tasks in order to make those work tasks and processes flow smoothly (p. 385). Star elaborates by describing such articulation work as the invisible "process of assemblage, the delicate complex weaving together of desktop resources, organizational routines, running memory of complicated task queues" that goes on below the surface of visible production work, enabling and supporting it (p. 387). She further notes: "This system is necessarily fragile (as it is in real time), depending on local and situated contingencies, and requires a great deal of street smarts to pull off. Small disruptions in the articulating processes may ramify throughout the workflow of the user, causing the seemingly small anomaly or extra gesture to have a far greater impact than a rational user-meets-terminal model would suggest" (p. 387). Star's focus is broader than memory practices, but the type of fragile and impromptu practices of articulation work that she describes here are, in fact, memory practices, the fragile and fleeting assemblages by which we continuously attempt to preserve our standing wave of memory through space and time. Further, many of these memory processes or devices do *rhetorical* work because they support and partake in the activities of invention in which we utilize contextual affordances to retrieve and manipulate stored information while communicating with others.

The rhetorical tradition, at least as it was originally formulated under the Greeks and Romans and based largely on the work of Aristotle, understood the acquisition of knowledge as fundamentally psychosomatic, involving the actions of both the mind and the body. In other words, such knowledge-acquiring practices represent articulation work, in Star's formulation. However, if we want to distinguish rhetorical memory practices from other forms of constrained or mediated articulation work, we must pay attention to memory practices related to the acquisition and use of knowledge. To achieve this, it is important that, like Star, we begin by paying close attention to the embodied aspects of human interactions with infrastructures in space and time. The rhetorical tradition points to the fact that only by accounting for the activities of memory performed by body and mind in tandem and in their situated environments can we fully understand these activities as culturally determined practices, as manifestations of particular memory regimes. Again, it proves instructive to begin with Aristotle when attempting to discern the taxonomy of memory practices that constitute the "street smarts" (Star, 1999, p. 387) of information managers working on "open-ended design processes[es]" (Sharples, 1996, p. 127) entailing composing and writing for the contemporary workplace.

ARISTOTLE'S EMBODIED EPISTEMOLOGY

While Aristotle does not offer the sort of tidy and quotable summation of the links among the body, composing, and memory that Plato provides in his famous critique of writing near the end of *Phaedrus*, his thought is indispensable for understanding memory practices as embodied activities. Although his theories about the body, thinking, language, and memory are dispersed throughout his works, Aristotle principally articulates them in the treatise "On Memory and Reminiscence," which Murphy (2002, p. 213) points out provides part of the "general theory of human action" that undergirds Aristotle's theory of rhetoric. First, it must be noted that Aristotle and his medieval successors like Aquinas believed that the body and its sense perceptions played a far greater role in thinking than has post-Cartesian thought. Theirs was a fundamentally embodied epistemology. That is, Aristotelians like Aquinas held that "the activity of thinking and the activity of having a sense perception are fundamentally analogous, not fundamentally different" (Carruthers, 1990, p. 57). Or, as Lakoff and Johnson (1999, p. 374) articulate it, Aristotle "locates reality ultimately in the world, and he thus sees our thought as dependent upon the nature of the world. . . . Thus, for both Plato and Aristotle, there is no separation between the mind and the world."

As for the role of memory in thinking, for Aristotle this too involves the whole body. In one of the most important but confusing parts of "On Memory and Reminiscence," Aristotle states:

One might be puzzled how, when the affection is present but the thing is absent, what is not present is ever remembered. For it is clear that one must think of the affection, which is produced by means of perception in the soul and in that part of the body which contains the soul, as being like a sort of picture, the having of which we say is memory. For the change that occurs marks in a sort of imprint, as it were, of the sense-image, as people do who seal things with signet rings. (Sorabji, 1972, p. 50)

At first glance, Aristotle seems to be saying that we only remember in pictures, which, if true, would, as Virginia Allen (1993, p. 51) succinctly notes, be a faulty notion, because "hasty introspection reveals that our knowledge of such things as typing, playing the guitar, and driving a car are not mediated with images."

Yet Allen misses a key nuance here. Aristotle is not saying that these "impressions" are exclusively or literally pictures or visual images, but, rather, they are "*like a sort of* picture" (Sorabji, 1972, p. 50, emphasis mine). Carruthers and Sorabji avoid Allen's mistake by labeling these "quasi imprints" *phantasms* to distinguish them from literal visual images (Carruthers, 1990, p. 16; Sorabji, 1972, p. 14). Carruthers (1990) glosses the phantasm this way: the "phantasm is the final product of the entire process of sense perception, whether its origin be visual or auditory, tactile or olfactory. Every sort of sense perception ends up in the form of a phantasm in memory" (p. 17). In other words, rather than exclusively (or even mostly) a visual picture, a memory is more like a multimodal "snapshot" derived from all our embodied senses in a given moment in space and time: "All mnemonic advice stresses the benefits to be gained from forming memories as 'scenes' that include personal associations . . . the need to impress the circumstances during which something was memorized . . . how one feels, the gestures and appearances of one's teacher, the appearance of the manuscript page, and so on" (p. 60). This interpretation of the Aristotelian phantasm as a "scene" perceived from the embodied perspective of the rememberer is seconded by Sorabji (1972, p. 7), who says, "Aristotle seems to imply [that] that the memory-image is a copy of one's view of that scene." Murphy (2002, p. 218) concurs, adding that data from the other senses are "collated" in the phantasm.

The process of deliberately recollecting memories (as opposed to simple random recall), then, entails finding or locating these phantasms, scenes, or snapshots, either via repeating some aspect of the physical circumstances in which the original sense impression occurred (e.g., walking to the foyer and retracing your steps to try to figure out what you did with your keys when you walked into your house) or by using some sort of artificial heuristic technique like the *loci/imagines* (place/image) mnemonic. The recollection process also involves an act of reconstructing and interpreting the embodied sensations laid down during the original experience of the thing being recalled: "recollection was understood to be a re-enactment of experience which involves cogitation and judgment, imagination and emotion" (Carruthers, 1990, p. 60). Since, in most cases we can't actually recreate the exact physical circumstances of the original experience (the memory work involved in most symbolic analytic tasks is rarely as simple as finding one's keys), artificial methods (that is, methods that are susceptible to training or the product of an art) and tools for recollection become vital.

Yet recollection during open-ended design tasks like composing and writing can be difficult and can resemble the cognitively demanding, labor-intensive process of dredging up one by one declarative memories that experts know how to avoid. To make the process of recollection easier, Aristotle offers a number of possible methods for tapping into and manipulating phantasms for purposes of deliberately recollecting memories during composing tasks: employing the *topoi* as "organizing modes of recollection" helps us to envision ourselves in relation to the points we are trying to make by "initiat[ing] memory in certain directions" (Murphy, 2002, p. 220); using metaphor, as I've already mentioned, helps us connect "concrete domains" to abstract ones (Lakoff & Johnson, 1999, p. 155); "tagging memory emotionally" makes remembering easier by adding an internal sensations to the external physical sensations of the phantasm snapshot (Carruthers, 1990, p. 60); and, perhaps most importantly—and as I have already alluded to in the discussion of hexis—repeating and practicing a particular composing activity habituates us to the memory demands of particular rhetorical situations (Murphy, 2001, 2002, points out that the Romans made habit the foundation of their educational system).

THE ARISTOTELIAN PHANTASM AND SITUATED THEORIES OF COGNITION

Each of the methods for engaging memory during composing that Aristotle offers makes intuitive sense as means of performing memory work, but how does the phantasm fare as a construct in light of contemporary developments in cognitive psychology, and, more pertinently, how might we study similar methods in the actual practices of contemporary writers as they work? Answers to these questions can be found in recent theories of social cognition deriving from the work of Vygotsky (1978) and Leontiev (1978).[2] In particular, theories of situated cognition, which posit that the activities and contexts in which learning occurs are inseparable from and coproductive of knowledge itself, prove especially relevant. First, consider the following passage, in which Brown, Collins, and Duguid (1989, p. 36) articulate the central construct "knowledge" lying behind their theories about how people learn in and through situated activities:

> Knowledge, we suggest, similarly indexes the situation in which it arises and is used. The embedding circumstances efficiently provide essential parts of its structure and meaning. So knowledge, which comes coded by and connected to the activity and environment in which it is developed,

is spread across its component parts, some of which are in the mind and some in the world much as the final picture on a jigsaw is spread across its component pieces.

It is not too much of an overstatement to suggest that the construct that Brown, Collins, and Duguid are describing here serves essentially the same purpose in remembering and thinking as the Aristotelian phantasm. That is, their "final picture," like the phantasm, is something similar to an embodied snapshot linking interior phenomena—knowledge—inextricably to the external circumstances of place and time in which that knowledge was learned—in which it entered memory (Brown, Collins, & Duguid, 1989, p. 36).

Subsequent theorizers of situated cognition add nuance to this construct. For instance, Reynolds, Sinatra, and Jetton (1996, p. 100) describe situated cognition in this way: "Situated cognition . . . attempts to account for how one learns in a conceptual environment. The conceptual environment consists of the external world as perceived, the internal representations of the perceptions, and the resulting interactions." These "internal representations" seem quite similar in both origination and in function to the phantasm. Cybernetic theorists, who have turned to situated cognition as a means of formulating new approaches to artificial intelligence, add further weight to this claim. For example, Clancey (1997, p. 5) says, "Conceptual knowledge, as a capacity to coordinate and sequence behavior, is inherently formed as part of and through physical performances. The formation of perceptual categorizations and their coupling to concepts provides material for reasoning (inference), which then changes where we look and what we are able to find." This "coupling" of perceptual information with declarative conceptual knowledge that Clancey recognizes, again, seems to function in much the same way that the phantasm does by collating diverse sensory inputs into usable and recollectable scenes which drive human thinking.

Moreover, some of the methods of tapping into and manipulating these "final pictures" or "internal representations" that the situated cognitivists posit as critical to achieving expertise in a given task domain resemble the advice offered by Aristotle for tapping into the phantasm. Habit, in particular, plays a critical role for both Aristotle and the situated cognitivists. According to Aristotle (1952, p. 693), "Acts of recollection, as they occur in experience, are due to the fact that one movement has by nature another that succeeds it in regular order. If this order be necessary, whenever a subject experiences the former of two movements thus connected, it will (invariably) experience the latter." Murphy (2002, p. 218) points out that what

Aristotle is describing is habit, noting that "the tendency to act in a certain manner, derives from memory in that unrecollected choices create a potential motion of the soul in advance of recollection."

An easy, if somewhat clichéd example of the way that cognitive knowledge or internal representations become embodied, physical memory practices is to look at sports athletes such as a professional basketball player who repeatedly practices his repertoire of shots until they become rote, automatic. Virtually everything becomes an integral part of the memory practice of taking a specific shot—the mental cognition and concentration, the player's vision, the timing and the balance, the physical movements required to get into position and then launch the shot—even the resilient feel of the ball and the bounce of the floor are assimilated into the fabric of the memory practice. One might even go so far as to argue that the sound of the ball on the hardwood and the smell of the arena might become integral part of the experience for the player, thus also playing something of an unconscious role in the broad scope of the memory practice.

The cognitive and embodied memory practice of the basketball player described in the preceding paragraph represents a hexis, as articulated earlier. This hexis is a mastery that results largely from the blending of cognition and memory with embodied practice within a given environment, or the very definition of situated cognition. As Hawhee (2004, p. 58) states, "hexis equals thought. Thought does not just happen within the body, it happens as the body." Or as Saugstad (2002, p. 385) puts it, for Aristotle, "the correspondence between the intellectual and the activity indicates that knowledge . . . emerges as a competence, where knowledge is a part of the person."

To deliberately trigger these "unrecollected choices" during some deliberate task like composing, then, one needs to practice so that the action becomes habitual during subsequent performances of the task: "Accordingly, therefore, when one wishes to recollect, this is what he will do: he will try to obtain a beginning of movement whose sequel shall be the movement which he desires to reawaken" (Aristotle, 1952, p. 693). Analogously, think of a person—a carpenter, for example—who failing to quickly apprehend, cognitively, the name of a specific tool that he needs to ask his apprentice to hand up to him ("It's on the tip of my tongue!"), starts impulsively to imitate with his hand the motion one would follow to use that particular tool, such as a hammering motion. Or further, as Carruthers (1990) glosses Aristotle's notion of habit articulated in the *Nicomachean Ethics*, "One's *hexis* or *habitus* is developed by the repetition of particular emotional responses or acts performed in the past and remembered, which then predispose it to the same response in the future. . . . Experience is made from many repeated memories, which in turn are permanent vestiges of sense perceptions" (p. 68). Carruthers links the role of the phantasm to the laying down of habit: "It is the spatial, somatic nature of memory-images that allows for secure recollective associations to be formed. . . . Because it is also a physiological process, recollection is subject to training and habituation in the manner of all physical activity (pp. 63–64). Finally, Sorabji (1972, p. 45) corroborates Carruthers by stating, "It looks as if Aristotle's view is that, whenever images regularly follow each other, this is by way of habit. The habit may have become established either because the images were naturally fitted to occur in a certain order, or . . . as a result of artifice [i.e., training]." Habit, of whatever type, is for Aristotle a product of physical as well as mental and emotional training.

Quite similarly, Brown, Collins, and Duguid (1989, p. 33) note that "understanding is developed through continued, situated use." From this idea, they formulate the notion of the "cognitive apprenticeship" as the ideal method of learning to be an expert in a given task domain. Again, as for Aristotle, the process of habituation achieved through extended apprenticeship is as much physical and emotional as it is mental:

> *Cognitive* emphasizes that apprenticeship techniques actually reach well beyond the physical skills usually associated with apprenticeship to the kinds of cognitive skills more normally associated with conventional schooling. This extension is not as incompatible with traditional apprenticeship as it may at first seem. The physical skills usually associated with apprenticeship embody important cognitive skills, if our argument for the inseparability of knowing and doing is correct. (Brown, Collins, & Duguid, 1989, p. 39)

So a cognitive apprenticeship is a means of achieving a *hexis*. The fundamental congruence of the Aristotelian conception of knowledge and that of the situated cognitivists leads them to articulate quite similar theories of learning.

Two other methods for manipulating the phantasm that Aristotle offers—the *topoi* and metaphor—suggest yet another set of cognitive theories rooted in the work of Vygotsky and Leontiev: distributed or joint cognition theory. Building on insights from Edwin Hutchins's seminal study of ship navigators in *Cognition in the Wild* (1995a), Engeström and Middleton (1996, p. 6) identify the central construct of knowledge lying behind theories of joint cognition in this way: "[The] unit of analysis [is] a culturally constituted functional group rather than an individual mind. This theory reconceptualizes 'information' as the propagation of representational states of mediating structures that make up the dynamic and substance of any complex system. These structures include internal as well as external knowledge representations, (knowledge, skills, tools, etc.)."

These information "structures" fill the same role in theories of joint cognition as phantasms do in Aristotle's theory of memory. Consequently, as this passage suggests, phantasms, like these information structures, exist not only inside the head of the individual but also in the collective, in the "functional group." That is, because of the shared spaces in which we live and work and the commonality of our embodied experiences in these spaces, the Aristotelian phantasm, as a snapshot of sensory experience, is inescapably both an internal and an external representation of knowledge. Viewed in light of joint cognition, Aristotle's phantasm is a theory of joint as well as individual memory because it offers an explanatory framework for collective activity, as the internal representations that constitute phantasms are propagated across individuals through language, through spaces, through tools, and through artifacts. Metaphors and topics, then, are methods not only for individual understanding and thinking but also for joint thinking, because they are methods for sharing and communicating knowledge representations.

To take another example: Describing their case study of joint cognition in airport workers, Goodwin and Goodwin (1996, p. 83) point out that "in these data we are able to catch a glimpse of the social and historical processes through which a community accumulates experience of the habitual scenes that constitute their working environment, and articulates for each other how these scenes should be properly interpreted." These "habitual scenes," then, give rise to the metaphors and determine the common topics or "places" which a community uses to recollect and to reason. As I note above, Murphy (2002, p. 220) points out that the *topoi* "initiate memory in certain directions." Metaphors and topics, then, depend on the "embeddness of knowledge"; that is, they assume that "the ability to see something is always tied to a particular position encompassing a range of phenomena including placement within a larger organization, a local task, and access to relevant material and cognitive tools" (Goodwin & Goodwin, 1996, p. 61). The airport workers are able to work together successfully because of the shared contexts in which their memories were originally laid down. By virtue of their training and common experience, they share phantasms, or, as Fentress and Wickham (1992, p. 59) put it, social memory is strongest "at the level of shared meanings and remembered images."

Finally, Vygotsky and Leontiev's explication of the concept of tool mediation helps us understand how the ancient rhetoricians may have employed Aristotle's embodied epistemology at the level of practice. Activity theory posits that tools operate at both external (embodied) and internal (psychological) levels, bridging the body and the mind in everyday practice: "Tools shape the way human beings interact with reality [and] the shaping of external activities eventually results in the shaping of internal ones" (Kaptelinin & Nardi, 2006, p. 70). Tools also function as enablers of joint cognition by bridging the variable knowledge of individuals: "Tools usually reflect the experience of other people who tried to solve similar problems earlier and invented or modified the tool to make it more efficient and effective. . . . The use of tools is an accumulation and transmission of social knowledge" (Kaptelinin & Nardi, 2006, p. 70).

Memory tools are, of course, no different from any other tool: they are socially acceptable aides to natural memory that often become internalized and are often employed in idiosyncratic ways by individuals during the course of actual practice. The function of a memory tool might then be thought of as an aide to maintaining the standing wave of memory that enables effective action in the present by providing just the right information about or from the past. The memory tool accomplishes this because it contributes to the sequential and jussive work of the memory practice. The loci/imagines mnemonic of the Roman rhetoricians was, of course, just such a tool—it enabled Roman rhetoricians to think on their feet in the Roman forum by helping preserve the order of their speeches and by ensuring that only the correct information would be "in sight" and therefore in mind at the correct time that it was needed. That is, by turning difficult declarative memory tasks like recalling the content of one's speech into easier procedural or natural memory tasks like walking through a memory palace, the loci/imagines mnemonic might be thought of as functioning as a sort of embodied simulation or a filmstrip version of Aristotle's phantasm, in which the orator imagines him- or herself interacting with places derived from habitual or imaginative experience. As Luria's (1987) case study in *The Mind of a Mnemonist* demonstrates, such mnemonics remain an effective tool for assisting memory, but, most importantly for the present discussion, it was originally theorized from an embodied epistemology that fully enlisted the body—its perceptions and sensations—in its memory practices. As a tool for memory, then, the loci/imagines mnemonic will probably remain relevant longer than many of the digital memory solutions detailed in the previous chapter's consideration of archival technology and office automation.

SUMMATION: MEMORY WORK AS RHETORICAL PRACTICE

As stated at the beginning, I set out in this chapter to articulate a working description of a rhetorical memory practice and to ground this description in rhetorical theory and contemporary psychology in order to arrive at an approach for studying such practices as components of rhetorical invention. In the course of the chapter, we have developed a framework for understanding memory work as rhetorical practice, first by acknowledging Bowker's (2005) description of the memory regime as a way of describing the collective memory practices of a given culture, and next by affirming, with Wenger, McDermott, and Snyder, that memory practices themselves represent "a set of socially defined ways of doing things in a specific domain: a set of common approaches and shared standards that create a basis for action, communication, problem solving, performance, and accountability" (Wenger, McDermott, & Snyder, 2002, p. 39). Next, we sought to define rhetorical memory practices through the framework of Aristotle's broader interpretations of both knowledge and practice as embodied phenomena originating in mind and body and sought to demonstrate that everything that an individual does or uses in service to a memory or information-managing practice—including the competent use of the tools, devices, and affordances of the modern office environment—is an extension of the mind and body, and hence the very definition of an embodied activity or practice.

Further employing the Aristotelian framework for learning and memory, I have argued that, as the case study chapters will demonstrate, the information managers at Software Unlimited make deliberate and measured use of individual and reasonably well-defined rhetorical memory practices in pursuit of both long- and short-term knowledge. Initially, in the short term, they employ memory practices principally to acquire technical know-how (techne). But, in the longer term, they also often employ these practices in pursuit of an embodied level of mastery, of hexis, and ultimately of the social, ethos-related mastery constituting phronesis. Finally, through appeal to Aristotle's embodied epistemology and the Aristotelian phantasm, I have argued that the rhetorical memory practices undertaken by these information workers are fundamentally *situated*, both cognitively and in terms of the environment in which they are performed, agreeing with Reynolds, Sinatra, and Jetton (1996, p. 100) that "situated cognition . . . attempts to account for how one learns in a conceptual environment. The conceptual environment consists of the external world as perceived, the internal representations of the perceptions, and the resulting interactions." In the next chapter, we will explore how this conception of rhetorical memory practice fits within the larger framework of communities-of-practice theory, principally following the work of Wenger and his colleagues.

However, for now, having thus defined, in general terms, rhetorical memory practices and located them within the conceptual framework in which they operate, we are ready to posit a taxonomy of the fundamental categories of memory practices specifically employed by the information developers at Software Unlimited as revealed by my research in this case study, as follows.

A TAXONOMY OF INFORMATION MANAGING MEMORY PRACTICES

During the course of the study, I assembled a large and heterogeneous body of data about the memory practices of the information developers at Software Unlimited, including field notes, video recordings, screen recordings, and textual artifacts. To produce consistent and coherent accounts of memory practices across these data, it was necessary to attempt to distinguish, as rigorously as possible, one type of practice from another. To accomplish this, I employed an iterative and modified grounded-theory approach to data coding and analysis. First, in an inductive phase, I employed a strategy of open coding (Strauss, 1987) to reveal provisional relationships among segmented pieces of data. In the second phase, I deductively drew from the extensive literature on information, knowledge, and memory that I summarized in chapter 1 and the insights offered by ancient and contemporary rhetorical theories of embodiment outlined in the present chapter to classify these data segments by a basic typology of activities or practices employed by individuals when managing information or doing memory work. This taxonomy included the following categories: archiving, reminding, finding, referencing, storytelling, and gesturing.

Archiving Practices

First, I categorized as *archiving* any practice the primary purpose of which appears to be storing information for a nonspecific or otherwise unspecified later use. I employ the term "archiving" rather than other possibilities like "storing" because of the richness of the term in the literature. Of particular importance in this regard is Derrida's (1998) definition of archiving practices as always containing both sequential and jussive elements: archiving practices are sequential because they usually entail standardizing and classifying information in some fashion so that it can always be found when needed, and they are jussive because they entail some judgment about which types of information should be kept and which types should be purged from memory.

The sequencing aspect of archiving practices corresponds in many ways to the metalevel activities of organizing that the personal information management literature identifies with activities like creating and maintaining file folder structures to support unspecified future activities. Since the potential future uses to which a piece of information may be put are not necessarily known beforehand, we attempt to make the information findable when needed to support a variety of activities. The following, then, are examples of archiving practices: maintaining a notebook containing notes taken at meetings, creating a directory structure on a hard drive, storing a document in a paper file folder, and copying a file to a network drive. Archiving practices are also, of necessity, jussive too because they also inevitably entail making judgments about which information is not worth preserving, as well as judgments about previously preserved information that is now obsolete and must be purged—something ever so common in today's rapidly changing technological world. The rationales for these decisions, many of which are made on the fly, arise from both individual experience or knowledge and the affordances and constraints imposed by the memory regime.

REMINDING PRACTICES

By contrast with archiving, I categorized as *reminding* any practice the purpose of which appears to be to create a cue to trigger a specific future behavior. Where archiving practices entail saving information for the long term and for multiple or not always clearly defined future uses, I distinguish reminding as creating a limited-term memory aid with a single, clearly defined purpose or a very limited number of purposes. Where archiving primarily facilitates conscious searching and browsing, reminding principally facilitates unconscious recognizing. The following are examples of practices that I characterize as reminding: creating a list of questions to ask at a meeting, creating a Post-it note and sticking it in a spot where it will be seen, and using the calendar feature of Microsoft Outlook (or some similar digital or physical affordance) to create a pop-up meeting reminder. Another notable feature of reminding practices is that, in most cases, they immediately become obsolete or "outdated" once they serve their intended function

Finding Practices

I categorize as *finding* practices in which an individual appears to be attempting to locate digital or nondigital information. For finding performed with computers, this principally includes the activities of conventional information retrieval: logical searching and location-based browsing, to use Barreau and Nardi's (1995) distinction. Whereas recognizing a reminder often occurs unintentionally, finding something stored in an archive is intentional (searching) or semi-intentional (browsing). The following are examples of practices that I label as finding: browsing through files on a hard drive, querying using the Find and Replace feature of Microsoft Word, rummaging around an office; searching the web, and looking for reading material in the company library.

Referencing Practices

I categorize as *referencing* those activities or practices in which knowledge held by another individual—or "housed" in document form, whether digitally or in print—is either consulted directly or referenced in communication. In one sense, then, referencing could be thought of as the activity that follows successful finding: after we have located a source document, we refer to it as we write as an aid to short-term memory, or we ask a colleague for help on a particular problem after we have found out that he or she possesses the piece of information of interest. It would seem that referencing may also involve appeal to authority, as might be the case in referencing a corporate policy document, or even a particular individual in a position of authority within the company, as warrants in an argument or debate. The following are examples of activities that I label as referencing: reading information from a whiteboard or a document, asking someone for information during composing, referring to knowledge or a type of knowledge held by another person during an interview, and consulting the user interface of an application while writing documentation for that application.

Storytelling Practices

A story is "a sort of natural container for memory," and the practice of storytelling is a rhetorical memory practice in every sense: storytelling interprets past events to meet current exigencies (Fentress & Wickham, 1992, p. 50). In other words, stories "offer a powerful means to understand what happened (the sequence of events) and why (the causes and effects of those events)" (Brown & Duguid, 2002, p. 106). And storytelling "can be effective for transferring both implicit knowledge about how things get done, as well as deeper tacit knowledge that reflects the values shaping behaviors" (De-Long, 2004, p. 102). Following Gabriel (2000), I categorized as *storytelling* any oral account of past events delivered in a meeting or interview that contains a substantive plot, distinguishable character(s), some form of action, and "a proper end" (Gabriel, 2000, p. 26). This method thus distinguishes storytelling from other forms that oral narrative takes in my research data, such as opinions, which do not contain plot, character, or action, factual reports, which merely convey information and "refus[e] to read any meaning in the events described," and "proto-narratives" or story fragments, with rudimentary plots and no clear end (Gabriel, 2000, p. 26).

GESTURING PRACTICES

Finally, borrowing from the gestural coding method advocated by McNeill (2005), if the participant performs a gesture high in the dimensions of iconicity (picture drawing) or *deixis* (pointing), I double coded the segment as *gesturing*. Research on gesture demonstrates that gestures, particularly spontaneous or noncommunicative gestures, are often outward indices of internal spatial and embodied memories: "Gestures are the person's memories and thoughts rendered visible. . . . The gesture reveals not only the speaker's memory image but also the particular point of view that he had taken toward it. . . . If we were to look only at the gesture or the speech, we would have an incomplete picture of the speaker's memory and mental representation of the scene" (McNeill, 1992, p. 12–14).

My rationale for paying attention to gestures is that they open a window onto an important type of information that would otherwise be imperceptible: the internal snapshots or images of previous experience that a person carries in memory. Gestures are, in other words, memory practices of precisely the type that Aristotle and contemporary situated-cognition theory tell us we should be interested in—they are visible manifestations of interior "snapshots" of experience, of Aristotelian phantasms, in other words; they make the past visible for us in the present, provided we are quick enough to spot them.

CHAPTER THREE

Studying Rhetorical Memory Practices in Context

Embodied literate activity is woven out of profoundly heterogeneous chains of acts, scenes, and actors oriented to diverse ends.
—Paul Prior & Jody Shipka, 2003, p. 230

S tudying memory is a complex undertaking. After all, any phenomenon that springs from human cognition encompassing epistemological and creative aspects of thought, and yet draws inferences from outward experience and environmental factors, is bound to be intricately complex and extremely difficult to dissect. In one sense, memory is ubiquitous, because memory as stored information or knowledge influences virtually everything we do, and, in practical terms within the workplace, the various aids to memory are legion, from our email in-boxes to reminders posted on our bulletin boards to our Internet browser bookmarks to the organization of papers on our desks. And of course, as we have established from our review of Aristotle in chapter 2, conceiving of knowledge as "stored information" in memory simply fails to do justice to either that knowledge or the memory practices through which we put it to practical use. After all, in order simply to function in the world, we are always in a conscious, constant state of memory use, or, to put it metaphorically, we are always immersed in the current of the "standing wave" of available memory. In another sense, of course, memory is ephemeral, often taking place inside our heads in seemingly random, fleeting moments that are a mix of deliberate recall, unintentional recollection, and even fantasy. Memory is often unpredictable: Anyone who has ever played along with the television show *Jeopardy!* has probably had the experience of getting the question to the clue right only to wonder where on earth that answer came from. Approaches to studying the function and effects of memory in human activities, such as writing, are similarly diverse, ranging from broad explorations of memory as social narrative in Fentress and Wickham (1992) to narrowly focused and quasi-experimental attempts to describe the cognitive processes of memory at work during various activities, such as the composing research of Flower and Hayes (1981) and Kellogg (1996).

Thus far in this book I have focused on memory work primarily as individual phenomena; that is, in endeavoring to understand memory work as rhetorical practice, I have focused on the strategies and tactics employed by individual information managers (as well as other workers) in order to do their jobs effectively and efficiently and to create new knowledge for their organization or company. I have argued that individual memory practices are the activities and tools by which members of a given memory regime attempt to deal with the stored, historical information or knowledge from the past while assimilating an ever-increasing, often formidable, influx of new information from a myriad of sources ranging from innumerable Internet sites to mass media, as well as from coworkers and other industry sources—even including information emanating from competitors in one's field. I have further argued that these activities and tools are both substantially influenced by the memory regimes in which they reside, but that they are also often employed in idiosyncratic ways in actual work processes based on the experience and the intelligence—both cunning and practical—of individuals as they respond to rhetorical situations, as we will see in the following case study chapters. We have looked at memory practices, however, principally from the point of view of the individual information manager, and somewhat irrespective of the larger context of the memory regime under which he or she must operate, or for that matter, of the overall corporate organization within which such information managers must do their jobs. We must now examine the memory work of individuals within and as a function or component of the larger context of the organizational structure and culture of the corporate entity—that is, the company itself.

At the very heart of the memory work undertaken or performed by the information developers at Software Unlimited is, of course, learning. On a very basic and practical, foundational level, as individuals who are hired as information developers, they must learn the company's product lines and their functionalities and features—presumably right down to the most intricate and subtle nuances. So, for example, they must learn about the new software innovations—whether entirely new products or simply upgrades to existing programs—that the software developers are inventing or working to develop and refine (as well as, for that matter, outmoded

learning

people w/in company

or obsolete features that the software developers may actually be replacing or removing from existing software packages in the company's product line). Furthermore, as basic employees simply trying to do a job and make a living, the information developers (just like any other employees) must learn the internal workings, procedures, and protocols that define and proscribe how the company functions and processes information and workflow organizationally—what might be described as the "nuts and bolts" standard operating procedures of the company: in other words, the "how," in answer to the question of just how the work gets done, or just how the company functions on a day-to-day basis. All of this information, or knowledge, or know-how, if you will, represents at least a portion of the overall memory regime of the organization; knowledge that certainly behooves new employees to learn quickly and fully, and, just as importantly, for longer-tenured employees to keep up to date with as things change, or as new innovation and organizational practices are—inevitably—introduced and adopted by the company and its growing staff of workers.

Because, of course, as we noted in chapter 1, we may reasonably expect that ambitious workers who wish to have successful careers realize that in order to achieve career success, they must perform their jobs exceedingly well as a means of helping their company grow and prosper, and to be profitable enough to offer greater salaries and professional advancement. To this end, of course, it is critical to learn how the company "does things"—in essence, to learn the corporate culture which is itself embodied in the company's overall memory regime. And yet, as I have argued, and as the literature has shown, workers exhibit a strong desire and demonstrated tendency to devise and employ their own unique sets of memory practices in order to succeed in (and be recognized for) their job performance, thus, ostensibly, helping the company to succeed as well. In order to achieve success in their job performances, typically workers must create, adopt, or adapt reliable memory practices that support not only the goals of the organization, but also its standards of operation. That, however, is not always the case, and we may look to one of the most famous firings in American corporate history—that of Steve Jobs from Apple in 1985—to see a breathtaking example of this. Based on recent remarks made by former Apple CEO John Sculley at the thirteenth annual Forbes Global CEO Conference held in Bali in September of 2013, we may conclude that Jobs at that early stage of his career failed to understand the differences between the privately held, highly entrepreneurial venture that Apple had been—and which could, and would, take outrageous risks—and the publically traded company it had become—which was accountable for its actions to investors and millions of shareholders (Lane, 2013). While it is certainly true that going public significantly and profoundly changed the corporate structure and the very culture of Apple, the relevant point here is that now, Job's free-ranging entrepreneurial management style violated the new goals of the organization as well as its standards of operation.

On a less celebrated level, as I noted in chapter 1, while the memory practices of individual workers are both guided by and enacted in direct support of the prevailing memory regime (or regimes) of the organization, such regimes often allow for a large measure of individual improvisation, preference, creativity, choice, and adaptation of memory practices among individual workers—as long as those practices ultimately further the company's prosperity and goals. Indeed, unique, creative ideas or innovations that represent minor violations of the corporate culture, but which are highly successful, are often not only welcomed, but even celebrated in American business lore, such as Ray Kroc's decision to eschew waited tables for fast food, leading to the vast empire of MacDonald's.

LEARNING AND COMMUNITIES OF PRACTICE

Yet none of this learning, or, more particularly for our purposes, none of the knowledge that accrues from all of this learning does the individual information developer any good on its own, or "in a vacuum," metaphorically speaking. As Wenger (1998, p. 10) points out, "Information stored in explicit ways is only a small part of knowing." Nor, especially, does any of this learning take place in a vacuum: Echoing Saugstad's (2002) assertion that Aristotle's broader view of knowledge is that knowledge is in fact "human activity," Wenger (1998) states further that "knowing involves primarily active participation in social communities" (p. 10) and describes participation as "the social experience of living in the world in terms of membership in social communities and active involvement in social enterprises" (pp. 55–56). Wenger proposes the concept of "communities of practice" to attempt to explain how learning as participation "reproduces and transforms the social structure in which it takes place" (p. 13). That "social structure" for Wenger is the community of practice, which he describes as follows: Wenger begins by asserting that the term "communities of practice" is "not a synonym for group, team, or network" (p. 74), but, rather, he argues that communities of practice "can be thought of as shared histories of learning" (p. 86). I think it is important to stress right from the outset that these "shared histories of learning" among individual members of the community are, of course, not exact replications; while there are important reference points of cohesion, there are also inevitably points of contradiction and even of conflict.

In any case, Wenger goes on to identify what he calls the dimensions of practice as a property of a community as

1. Mutual engagement
2. Joint enterprise
3. Shared repertoire—materiality

Mutual engagement, according to Wenger, involves not only the competence of the individual, but also the competence of others in the community. In any social community there are aspects of overlapping (and reinforcing) expertise among members, but there are also aspects of complementarity: It is not unusual, for example, for members of a mutually engaged community to divvy up the tasks confronting it according to the simple logic of "who's good at what." This notion of mutual engagement then leads to the second dimension of joint enterprise in which there are indigenous tasks which all members of the community must perform, while at the same time there is room for negotiated assignment of tasks that, as just noted, might be delegated to specific members of the community based on individual expertise, facility, or efficiency, or even on individual preference or affinity for a particular task. But the important aspect of joint enterprise in a community of practice is what Wenger denotes as the creation of a regime of mutual accountability.

Generally speaking, a well-defined community of practice is keenly aware of its need to perform not just competently but expertly in the eyes of other, related communities, and especially in the eyes of the larger organization (which we might practically read as "looking good in the eyes of management"). This sense is particularly acute when the community of practice is most strictly delineated in stark, easily identifiable or highly visible terms, such as a specific department of a company established to perform a very specific menu of tasks. And as we will see, this is very much the case with the members of the Information Development Department, who, at various points in this study, are at pains not simply to present a united front to other communities within the Software Unlimited galaxy, but also to argue and strongly advocate for having their say on matters of performance and policy which the information developers collectively believe to be their responsibility, or "turf." Thus, it is important to note that while this regime of mutual accountability may lead, positively, to some degree of efficient task delegation based on which members of the group can most effectively accomplish specific tasks (thereby advancing both the goals and the perceived expertise of the community, or department, in this case), mutual accountability may also require some degree of conformity to a particular reading of the community's "shared history," which may in turn actually restrict or curtail the group's effectiveness, and thereby actually create an impression directly opposite the one desired by the community. That is, such actions requiring member allegiance or conformity to a constricted version of its shared history may lead to a diminished perception of the group's expertise, its ability to be creative and innovative, and in particular its ability to function as a team with other communities under the general corporate culture and its inclusive memory regime: a dangerous combination, particularly in a company based on and ostensibly driven by the very concepts of rapid innovation, flexibility, and active participation and interplay among individuals and teams.

Wenger describes the third dimension of practice as the property of a community, "shared repertoire—materiality," as "routines, words, tools, ways of doing things, stories, gestures, symbols, genres, actions, or concepts that the community has produced or adopted in the course of its existence, and which have become part of its practice" (1998, p. 83). Significantly, the terms and phrases that Wenger uses evoke many of the memory practices that I identified among the information managers in this study, as delineated in chapter 2. For Wenger, these shared resources are the community's "repertoire of practice [a repertoire that] reflects the community's history of mutual engagement and its shared histories, but remains inherently ambiguous" (p. 83). As Wenger states, "Histories of interpretation create shared points of reference, but they do not impose meaning. Things like words, artifacts, gestures, and routines are useful not only because they are recognizable in their relation to a history of mutual engagement, but also because they can be reengaged in new situations" (p. 83).

In communities of practice, then, for Wenger, practices evolve over time as shared histories of learning. "History in this sense," Wenger explains, "is neither merely a personal or collective experience, nor just a set of enduring artifacts and institutions, but a combination of participation and reification intertwined over time" (1998, p. 87). Now, it must be noted that Wenger's conception of a community of practice does allow for some discernible measure of the contradictory or conflictive aspects, which he terms "discontinuities," which I assert exist within such communities, as I alluded to at the beginning of this section. As Wenger puts it, "Because a community of practice is a system of related forms of participation, discontinuities propagate through it. When newcomers join a community of practice, generational discontinuities spread through multiple levels; relations shift in a cascading process. Relative newcomers become relative old-timers. . . . Participants forge new identities from their new perspectives" (p. 90).

All of this is well and good, and Wenger's description of how newly hired individuals become indoctrinated or familiarized with the shared histories and the operational workings (the shared repertoire of practice) of a community or of a corporate department, as the case may be, strikes us as comfortably and intuitively accurate: Certainly, newcomers "learn the ropes" and eventually become seasoned old-timers. But Wenger's explanation for the appearance and persistence of such discontinuities does not go far enough. To address this gap, we need to examine the process or criteria through which, in Wenger's view, a newcomer to a community of practice—whether one with no prior experience whatsoever or one with an established work history with, say, a competitor firm—may successfully gain entry to becoming a full-fledged member of that particular community. Wenger (1998) warns that "membership in a community of practice is not something that can be granted arbitrarily" (p. 136). In fact, Wenger stresses that the competence required for membership is not simply the paradigm "job description" variety of skillset that, technically at least, nominally qualifies the newcomer for the position: "This [required] competence is not merely the ability to perform certain actions, the possession of certain pieces of information, or the mastery of certain skills in the abstract" (p. 136). These are all qualities that we would expect in the seasoned newcomer with previous industry experience, yet, as Wenger rightly points out, these alone are not enough to gain immediate, or even a fast-track road to, acceptance as a member in good standing with the community of practice.

Rather, Wenger (1998, p. 137) asserts that membership is only granted after the newcomer has assimilated the three dimensions of a practice that we discussed earlier: "mutuality of engagement, accountability to the enterprise, and negotiability of the repertoire." Here again, all of this makes a great deal of sense. Remember that in the beginning of this chapter I made the point that at the very heart of the memory work that individuals must do in their workplace is learning and that an integral and obviously very important component of that process is learning specifically "how the company works." As a functionality of the larger organizational regime of the company itself, by extension it makes perfect sense for the newcomer to learn how his or her community(ies) of practice work. The oftentimes pejoratively indicated phrase "That's not how we do things here at XYZ Corporation" can, in fact, mean something extremely positive, as when used to explain, for example, that XYZ Corporation puts a high premium on cus-

tomer service and product satisfaction. In a sense, then, even a highly experienced newcomer must undergo an apprenticeship of sorts, however brief
that may or may not be.

So where does all of this set the bar for admission to a community of
practice? Wenger (1998, p. 137) explains, "It is by its very practice—not by
other criteria—that a community establishes what it is to be a competent
participant, an outsider, or somewhere in between. . . . A community of
practice acts as a locally negotiated regime of competence." What Wenger
seems to be saying here, and I believe accurately, is that communities of
practice tend to set standards of conduct for their participant members to
follow: essentially and for the most part, one must choose from the community's "store" or "shared repertoire" of practices in performing one's
work-related tasks. Perhaps unwittingly, a community of practice may in
this way act to impose a certain measure of conformity on its participants
much in the manner I described earlier in discussing the second dimension
of practice, joint enterprise with *mutual accountability*. However, the concept of community of practice in my view does not go far enough in describing or accounting for individual creativity and innovation in the workplace.
The approach to studying memory work presented in this book represents
a refinement and an enhancement of the conceptions of communities of
practice and of situated cognition that provides more power to explain the
broader aspects of learning which those two conceptions do not. My research indicates that a primary focus on rhetorical memory practices with a
consideration of the memory regimes within which they reside helps us to
explore learning in more detail and in multiple modalities: Learning is embodied and mental, tacit and explicit, individual and social. Communities-
of-practice theory speaks to aspects of some of these in part, but it does not
focus on "memory practices" specifically, and therefore it misses the ways
in which individual members of communities of practice deliberately, but
not always fully consciously, attempt to integrate with their communities
of practice using those very memory practices. Moreover, communities-of-
practice theory does not focus on those independently creative, idiosyncratic, and unorthodox instances wherein some individuals innovate by
breaking with the established repertoire of practices acknowledged by or
available to that community. As Fox (2000, p. 860) notes, communities-of-
practice theory "tells us little about how, in concrete practice, members of
a community of practice change that practice or innovate," a critical factor
that we discuss in greater detail later in this chapter. By contrast, the research reported here strongly indicates that a careful exploration of rhetorical memory practices can and does offer valuable insights into individual

improvisation as well as it does into other workplace activities. In sum, and in relation to this research study in particular, studying the rhetorical memory practices of the information managers at Software Unlimited enables us to explore, I assert, the entire range of strategies and tactics used by these workers to manage information overload and to create and share new knowledge for the benefit of the entire company. Next we turn to how we may go about analyzing individual rhetorical memory practices in the context of the communities of practice in which they are embedded.

ANALYZING MEMORY PRACTICES IN CONTEXT

In chapter 1, I introduced Bowker's (2005) concept of the memory regime as a useful shorthand way of describing the collective memory practices of an organization or group. Bowker describes the process by which memory regimes form—that is, the process by which collective practices become regularized or standardized—in this way: "Memory regimes . . . articulate technologies and practices into relatively historically constant sets of memory practices"—standard ways of thinking about memory and doing memory work (p. 9). Although Bowker may not be deliberately making the connection, I would suggest that Bowker's use of the term "articulate" here leads us naturally to consider a strand of social theory and a particular set of research approaches that focus on the dynamic relationship between individual practices and the social and embodied contexts in which these practices are embedded.

Articulation theory holds that the relationship between the larger context and individual practices is variable and dynamic rather than fixed and static. In other words, it holds that context and culture do not strictly determine individual practices, but rather that individuals and groups routinely exercise ingenuity to adopt their own or to appropriate and adapt existing practices, tools, spaces, and relationships to further their own long- and short-term interests and purposes (Hall, 1996). Like the rules of syntax in a language, which cue expectations about the order in which we articulate sentences but do not determine what those sentences contain, the norms, ideals, or expectations of a given culture or regime do not entirely prescribe the actual manifestations of those norms or ideals in the real world of human activity (see Slack, Miller, & Doak, 2004).

By invoking articulation in formulating the concept of the memory regime, Bowker offers rich possibilities for analyzing the components of rhetorical memory practices as a mix of influence and improvisation. However, Bowker is more interested in the historical development of scientific disciplines over long periods of time than in the dynamic interplay of cultures and practices that occur in the world of everyday activity. As a result, the full analytic and interpretive possibilities offered by the concept of the memory regime go undeveloped in Bowker's own work. In this book, I too am less interested in exploring the concept of the memory regime than I am in examining individual memory practices, which nonetheless fall under the auspices of an overall corporate memory regime. However, to derive an approach for researching and interpreting *situated* practices (i.e., social and embodied in real situations) in workplaces, we first need to unpack the concept of the memory regime and the relationship between cultures and practices that the term signals. In the following section I attempt to do this through a brief general discussion of the challenges to studying individual memory practices within the framework of the overarching memory regime and by then examining specifically the practical, organizational structure of a typical—and what might be described as a traditional—corporate memory regime, followed by a similar analysis of the more atypical and unorthodox memory regime that exists at Software Unlimited at the time of this study.

CHALLENGES TO STUDYING MEMORY REGIMES

In chapter 1, I noted that the memory regime of an organization both cues and constrains the memory practices that take place within it by attempting to specify, among other things, which information is worth preserving, what counts as useful knowledge, who is empowered to do the memory work of turning information into knowledge, what tools and other affordances are available for doing this work, and who has access to these tools. I also noted that the memory regime is both constructed from and manifested in the built, social, and software infrastructures of an organization, including things as varied as the database technologies it uses, the software tools it provides, the war stories its members tell, and the information-retention policies dictated in its policy manuals. Some manifestations of the regime are fairly obvious and easy to spot (e.g., an organization that has a long-standing contract with Microsoft requires its employees to use Outlook to keep and share their calendars. Such is the case at Software Unlimited). Other manifestations of the regime, however, might not be as easily discerned or as straightforward as they first appear (e.g., again as at Software Unlimited, despite the contract with Microsoft, one particular work team, and that team only, uses a different groupware calendar. When the researcher asks why, no one offers the same answer). How do researchers both see and interpret not only the obvious but also the many less-than-obvious elements that characterize the relationship of a local memory culture to the situated practices within that culture? As a first step toward answering this question, we might think of the research situation as a set of challenges.

First Challenge:
The Interplay of Improvisation and Constraint

As I mention earlier, the first and fundamental point that the concept of articulation signals about the relationship of memory regimes to memory practices is that, although the regime shapes practices, it does not determine them. Human beings are complex entities with different experiences that shape each person's own processes of memory. As a result, individuals bring to their current workplaces a wealth of previous experiences with managing information and creating knowledge from their personal lives, from their educational experiences, and from their previous jobs. Inevitably these individual experiences shape the ways in which people use the information tools, infrastructures, and stories provided, in general terms, by their current employer. In other words, the insights we derive from our previous experiences and from our ongoing situated activities offer us ways of performing tasks different from those offered or officially sanctioned by the organization. In a sense, then, we are always seeking to make our present external memory systems better correspond to our internal memories of past experience. Moreover, working under the assumption that most employees want to make a positive, substantive contribution to the organizations they work for, individuals whose "internal memories" are of very *successful* past experiences or practices are likely to want to introduce them into the external memory systems—the memory regime—of their current employer or organization. This would certainly appear to be true if such individuals are career minded, seeking rapid advancement, and therefore might be interested in "shaking things up" a bit. However, it also stands to reason that individuals who have discovered or devised unique methods, procedures, or techniques in the past that enabled them to perform their job tasks more efficiently and more easily are apt to want to bring those innovations along with them when they change employers (and hopefully share them with their new colleagues!).

The role that such improvisations play in cultural development is a major theme of a large body of cultural-studies research (e.g., Hall, 1996; de Certeau, 1984; Bourdieu, 1990). Studies of organizational or workplace learning informed by cultural-studies perspectives have long recognized modifications, alterations, clever emendations, and idiosyncratic employments of standard procedure as a recurring feature of workplace practice (Star, 1999; Orr, 1996; Brown & Duguid, 1991). In fact, Brown and Duguid (1991, pp. 41–43) usefully distinguish the difference between the standard, sanctioned way of doing things and the modified version as "canonical" versus "non-canonical" practices: canonical practices are what we do when we follow the instruction guide or policy manual to the letter; noncanonical practices are what we do to get the real job done when the official version proves inadequate.

Memory practices, though, appear to be a class of practice so susceptible to individual improvisation that, as I argue in this book, they are for all intents and purposes specifically designed for individual improvisation. This factor is demonstrated by both the research on situated memory practices and our own experiences. Perhaps the earliest thinkers who studied memory practices, the Greek and Roman rhetoricians who sought ways to help orators memorize their speeches, noted that all people do memory work a little bit differently based on their unique perspectives and experiences. For example, the *Rhetorica ad Herennium*, a rhetorical handbook for orators long attributed to Cicero, notes, "Things seem different to different persons [when creating visual mnemonics to help remember speeches]. Everybody, therefore, should, in equipping himself with images, suit his own convenience" ([Cicero], 1954, III, xxiii, 35–38). Consequently, when selecting mnemonic images to associate with the parts of a speech, the rhetor should choose images with vivid personal associations, because standard or generic images will not be as memorable.

Contemporary studies informed by cognitive psychology, including situated- and joint-cognition research, validate these ancient observations. They do so through clinical and situated workplace research that demonstrates that people remember best when they can modify or change the material that they need to remember in order to best suit their own needs or preferences (e.g., Moè & De Beni, 2005; Bellazza & Buck, 1988; Cornoldi & De Beni, 1991; Kirsh, 1995; Brown, Collins, & Duguid, 1989). Interestingly, recent discoveries in neurocognitive research even offer a possible neurological explanation for this phenomenon. One team of researchers has reported that, in terms of brain processes, remembering is to a large degree the same as reliving an experience: the same set of neurons fires when we remember an experience that fired when we first had the experience (Gelbard-Sagiv et al., 2008). Therefore, people quickly learn that, if they want to remember something better, they must strive to make the memory as vivid and distinct from all other experiences as possible.

There is also ample evidence from our own daily experience to confirm what research has learned about the variability of people's memory practices. Consider, for example, the ways in which individuals organize their personal filing cabinets. The affordances of the hanging file folder, the shape and dimensions of the particular type of cabinet available in their workspace, and, perhaps, company policies on information retention often dictate what is kept in the cabinet. However, the actual organizational schemes manifested in the order, distribution, and naming conventions used on the folder labels may vary considerably from person to person based on his or her own past experiences or perceived future needs. From a high-tech perspective, this is perhaps why the file folder and desktop metaphors for computer workspaces have, despite their many limitations, remained popular: they are flexible and give us some sense of control over how we visually arrange our information (Barreau & Nardi, 1995).

Research and experience appear to agree, then, that memory practices are a particularly rich source of improvisation and variability—and, to be perfectly plain, idiosyncrasy. Given even a slight degree of freedom to do so, we do our memory work our own way in order to make the contents of our external memories more memorable by more closely aligning them to our internal memory. The result, as Bowker notes, is that there exists "a dazzling array of memory practices that we engage in on a daily basis" and that "there are few censuses of these practices" (2005, p. 9).

Second Challenge: Influence Flows Both Ways

The concept of articulation signals a second important aspect of the relationship between a memory regime and its practices: The path of influence does not flow in one direction only. That is, memory regimes not only shape and articulate practices; practices also shape and articulate memory regimes. Practices and regimes are thus coconstitutive or coarticulatory: like practices, regimes are never entirely stable but, rather, might best be thought of as snapshots of collective action, as processes rather than products, or as "effects of outcomes, rather than . . . explanatory resources" (Law, 2007, p. 157). I would further argue that the reason that practices and regimes are never stable is that knowledge itself is never stable or permanent. As Saugstad (2002, pp. 378–79) puts it, "Knowledge in the practical area . . . can never become certain. It is a 'doxa' knowledge, which means that it is a casuistic, experience-based knowledge of the possible and the probable."

This inherent duality of the relationship between practices and regimes makes clear the participatory aspect of individuals within an organization (or a community) that Wenger identifies as being so critically important.

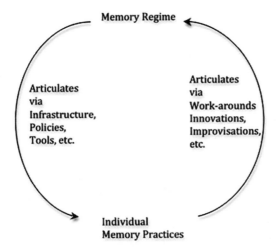

Figure 3.1   Cycle of innovation: An individual memory practice may be adopted by the larger organization and become a canonical practice of the memory regime.

And while I have been somewhat critical of what I perceive as the inability of communities-of-practice theory to deal adequately with individual improvisation and innovation, I believe that Wenger does recognize the underlying reason why creativity and improvisation in this context are so important to us. Wenger (1998, pp. 56–57) states that while "participation in social communities shapes our experience, . . . it also shapes those communities; the transformative potential goes both ways. . . . Our ability (or inability) to shape the practice of our communities is an important aspect of our experience of participation." What better sense of participation could one have than if one's improvisational and innovative practices served to positively influence coworkers or colleagues to modify their own practices?

This bidirectionality of influence is a result of the way in which innovations in practice circulate. First, if the improvised memory technique is effective, the individual's improvisation often does not remain private. Instead, the individual might share this innovation with coworkers in an attempt to help those coworkers perform a memory-related task, or the coworkers might accidentally observe or discover the innovation and try to apply it in order to better remember something (see fig. 3.1). Consider, for example, how often employees create acronyms, phrases, or slogans that help them and their colleagues remember a particular process or concept related to their work. This noncanonical practice might then be adopted in part or in whole by other individuals and work teams, and this ongoing process of continued adoption and adaptation might eventually become part of the standard way of doing things at the organization—a integral part of its overarching memory regime.

Cultural-studies theory is not alone in positing such a circular path of influence. Many recent theories of the evolution of organizational knowledge describe a flow in which emergent practices influence workplace cultures and in turn mold subsequent practices, thereby increasing the organization's knowledge (e.g., Nonaka & Konno, 1998; Brown & Duguid, 1991; Lave, 1991). And, informed by these theories, research on workplace practices offers abundant examples of the forms these innovations may take: they might result in the creation of new genres for sharing information (Orlikowski & Yates, 1994; Spinuzzi, 2008, 2003b); the development of standard work-arounds to limitations in software (Spinuzzi, 2003a); or new best practices for ensuring that important knowledge is not lost (DeLong, 2004).

It is this two-way flow of influence that holds the potential to make memory practices so valuable to professional and technical communicators as information managers in their workplaces. As in other domains of practice, improvisations in memory practices often represent far more than stubbornness on the part of an individual worker. They also frequently reveal the cunning intelligence of the worker in solving problems using the knowledge and resources at hand (de Certeau, 1984). We often modify standard ways of remembering not only so that we can find information more quickly or easily, but also to help ourselves understand and interpret it. In other words, memory practices such as creating a unique filing structure or a new spreadsheet can help us think about or visualize problems, to see connections between information, and to draw inferences and conclusions. Therefore, because technical communicators not only manage information for themselves but also help manage and interpret it for their teams and organizations, an innovation introduced to help with individual memory processes may hold considerable value for others as well, thanks in part to the regularities imposed by the memory regime itself. This interpretive and creative aspect of memory practices is, as I mentioned in chapter 1, the reason I designate these "memory practices" rather than "information management activities."

THIRD CHALLENGE: GRANTING POLITICS THEIR DUE

The third aspect of the relationship between a memory regime and memory practices that articulation emphasizes is that the relationship develops and unfolds within a specific set of organizational politics, history, and power relations: "Articulations . . . always have to do with power, the ability or inability to achieve effects, and they always involve relations of agency" (Slack & Wise, 2005, p. 174). This is, to a certain extent, stating the obvious. The political nature of memory is, of course, implicit in the term "memory regime." More importantly, we certainly do not need to be told that our workplace agency—our ability to perform or modify work practices, to "make things happen"—is shaped and constrained not only by our skills and abilities or by the affordances of our tools, but also by factors like our job title, our reputation, our seniority, the chain of command, the structure of the organization, our relationship with our teammates, or, most frustratingly, by unwritten or unspoken rules of unclear origin (Slack & Wise, 2005, p. 174). These are all, to a large extent, considerations of organizational politics and power. However, the precise role that these considerations play in cuing or constraining our situated practices can be both difficult to see and difficult to interpret.

The difficulty of accounting for the effects of power and politics on the small-scale or everyday activities through which people interact with information is not only a problem for busy professionals attempting to better understand their own workplace cultures. It is also a factor affecting more formal research approaches. In fact, failure to adequately account for the effects of political and power relations on individual memory practices has been noted as a weakness in several approaches to studying information and knowledge in organizations and communities of practice. Duguid (2005), for instance, notes that, in some cases, this failure may result because an approach focuses too much on organizational or community dynamics and too little on situated practices. In yet other cases, a criticism has been that because the approach is so preoccupied with the inner structure of individual activities it ignores issues of power (e.g., Fox, 2000; Blackler & McDonald, 2000). However, articulation implies that every aspect of the relationship between individual identity and agency and the social groups in which they are embedded is provisional, dynamic, and, therefore, negotiated. Thus, we must attempt somehow to grant situated activities and group dynamics the appropriate weights in our accounts.

To summarize, the principal challenges to studying the relationship between memory regimes and memory practices are the following:

> Memory regimes influence but do not entirely determine the memory practices that occur within them. Individuals and groups routinely modify memory practices to suit their situated needs for remembering, processing, and communicating information. The memory regime is emergent and can only be seen in practice.

The memory regime may in turn be shaped by these individual memory practices. What appears to be a novel innovation today may become the accepted method for doing things tomorrow.

Organizational politics and power relations always affect this two-way relationship between memory regimes and their memory practices. Agency to participate in or modify a practice is not evenly distributed across an organization.

One way to conceptualize these aspects of the relationship between a memory regime and its memory practices is as problems of scale in research. That is, they help illustrate the difficulty in determining how much weight to grant the regime and how much to grant individuals in processes of coarticulation. A memory regime is, in a sense, a moving target, a "slippery phenomenon": when and where can we see it and how much (if any) weight can we give it in our analyses (Law, 2007, p. 598)?

A PRACTICAL OVERVIEW OF THE MEMORY REGIME

It stands to reason that virtually all companies and organizations have, and indeed must have, memory regimes of one sort or another. The differences between the memory regimes of these organizations are largely a matter of degree in terms of how strictly such regimes—in the form of the reigning corporate culture—are applied, fostered, and indoctrinated, or "enforced" (if I may use that harsh-sounding term) by upper management and, conversely, how strictly the memory practices of the employees conform to the mandates of that larger corporate culture or memory regime. That is, companies may vary widely in their adherence or nonadherence to a strong, centralized, hierarchical, and preferred ideological and practical system of business practices.

For example, many sales organizations and telemarketers create very carefully crafted scripts for their sales representatives to use when dealing with prospective customers. These scripts are designed to maximize sales—the language is specifically constructed to encourage, even to induce prospective customers to say yes to an offer to buy a product or service. Similarly, most of the large retail department and big box stores provide their on-the-floor salespeople with scripts designed to help them get consumers to sign up for store-issued credit cards or to suggest add-on item sales at the register. As a memory practice, the issuance of highly formulaic verbal scripts to guide worker performance might go under the heading of storytelling practices, but what is more important to understand for my purposes

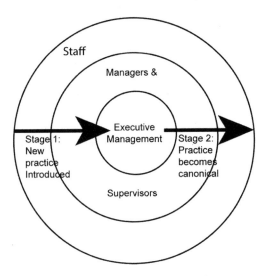

Figure 3.2   Spread of new memory practices in a traditional corporate memory regime.

here is that it represents a more rigid and top-down approach to an organizational memory regime and a more unified or strictly regimented corporate culture. Such organizations frequently believe—often with good reason based on their own sales performance data statistics—that these formulaic scripts enable their sales people to maximize purchases by their customers. Therefore, they are apt to strongly insist that their sales force strictly follow the scripts, even to the point of doing so verbatim. Innovation under such strictly defined protocols is not appreciated; rather, it is aggressively discouraged, and failure to comply with the preferred "message" can get an employee disciplined and even "written up"—or fired.

And indeed, it is perhaps this model of a corporate culture that most closely accords with what we think of when we use the term "regime," due mostly to its use in discussions of autocratic political organizations or governments. Thus, when we talk about a regime, we think of a centralized governmental authority with a uniform, and universal, often pinpoint message and mission.

Such a corporate culture might be represented, as shown in figure 3.2, as a series of concentric circles beginning with a strictly defined memory regime emanating from the executive management team in the middle, first to managers and supervisors in the next circle, and then to the rest of the company staff in the outer circle. To spread in this, arguably, more traditional corporate culture, an innovative memory practice formulated by an individual or a community of practice must pass up the chain of command from staff to supervisors to executive management, which then decides whether the innovation may be disseminated to other communities of practice.

The corporate culture at Software Unlimited, however, stands in stark contrast to this centralized authoritarian model. No doubt, as it is other cutting-edge software and technology companies, the creative innovation of new products, as well as upgrades to existing ones, is crucial to their success and growth, and thus individual creativity among the programmers, designers, and others within the organization is not only highly desired, but also strongly encouraged. In fact, the desire for individuals with high creativity and a certain irreverence for rules-based formulaic thinking are more than likely reflected in such companies' hiring literature and actual practices (as they are at Software Unlimited), through which they undoubtedly seek exceedingly "bright" people who can "think outside the box." The result is that the prevailing memory regime at Software Unlimited is both more diffuse and free-form and more dynamic in nature. That is, Software Unlimited's memory regime is for all practical purposes "spread out" more broadly among its employees, and rather than consisting of a centralized corporate theme emanating strictly and authoritatively from the uppermost echelons of executive management, the company's overall memory regime is far more changeable from day to day and may ultimately align more with the individual memory practices of its staff of programmers, designers, technical writers, marketing and promotions people, and trainees than with a company-wide paradigm. As I said earlier in this section, as a company organization, Software Unlimited certainly has a memory regime, but it would appear that it is significantly—and more loosely—disbursed across all of its departmental groups, as well as among the employees within those groups, with no one group possessing the definitive recipe for the corporate culture as a whole. In effect, there is no corporate bible at Software Unlimited. To help understand how this might work, figure 3.3 presents a diagram that may accurately reflect the relationships among five important groups (or departments) within the company relevant to this study:executive management, the software development team, the information development team, the training team, and trainees. (For a full exposition of company organization, the specific research sites and participants, as well as infrastructures, workspaces, and tools, see appendix A. Appendix A also provides a more detailed discussion of the research methodology employed for this case study.)

As can be seen in figure 3.3, while there is considerable overlap among the various teams, the teams do not share a universal, core, company-wide, and definitively articulated culture, and, further, there are aspects of each

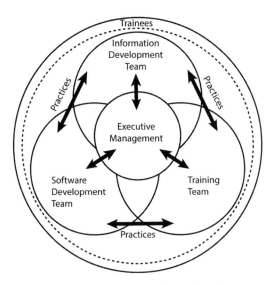

Figure 3.3   Spread of new memory practices in a decentralized corporate memory regime.

team's separate subculture that are also unique to that team and not shared even with the other teams.

The reason this dynamic aspect of the relationship between the overall memory regime and individual memory practices will be particularly important when we evaluate the memory practices at Software Unlimited is that, unlike the more conventional corporate model depicted in figure 3.2, in which presumably the preponderance of influence flows in top-down fashion from upper management down to managers, and only then to staff, or conversely, bottom up from staff to management in this bidirectional model, at Software Unlimited the dynamic appears to be very much different. What we have characterized as a two-way flow of influence within a more conventional corporate structure may in fact be multiplied by at least three (and perhaps several) spheres of influence operating within the overall corporate structure of Software Unlimited, as may be seen in the diagram in figure 3.3. Specifically, in some sense, and at a minimum, it would seem that upper management, and the software development, information development, and training teams at the company are all exerting influence on each other.

While this open atmosphere might be very positive for encouraging and promoting individual or group creativity and technical innovation, the danger lies in the potential for organizational fragmentation and the inadvertent creation of the "content silos" of information and knowledge that I alluded to in chapter 1, in spite of the use of various electronic groupware solutions and the attempted move toward an Agile development methodology, which was implemented specifically to ensure both intrateam and company-wide information sharing and access. On the one hand, sometimes innovations are implemented so fast that it is difficult for users to keep up; on the other hand, information-sharing systems are only as good as the people who use them and the extent to which they are able to keep shared information up to date.

Nevertheless, it would seem that the fast-paced, seemingly "overnight" innovation that is critical to the success of a cutting-edge software company makes this kind of multiple corporate culture structure a necessary evil, and there also appears to have been considerable concern on the part of Software Unlimited to proactively prevent the isolation of important information within discrete departments of the organization, which is evident in the information and reporting systems, both electronic and social, that the company has set up to ensure that information is shared effectively, efficiently, and as transparently as possible. We will also see later on that these efforts have not always been completely successful.

In this section I have presented a general, practical overview of the memory regime as a functionality of a corporate entity or similar organization. More particularly, I have attempted to paint an accurate and detailed picture of the comparatively unorthodox organizational structure of Software Unlimited, in which there is no central or universal memory regime emanating in top-down fashion from management through the echelons of employees, but, rather, there exist several constelled memory regimes corresponding loosely to distinct communities of practice, and further still in some instances running along departmental lines, yet coexisting and cooperating with one another reasonably, if not entirely seamlessly (or so one would hope). My hope is that this expository analysis of the formal corporate structure of Software Unlimited, such as it is, and especially what appears to be the less formal social structure through which the company's individuals and communities of practice—through which, I assert, the real work of the company gets done—will provide valuable insight into the rhetorical memory practices of the information developers who are the participants in this study, and whom I discuss in the following case study chapters.

As noted earlier, most of the hard-data information regarding the company itself, its organization and history, physical plant, and infrastructure, as well as its workspaces, equipment, and tools, are described in detail in appendix A. However, there are two issues of sufficient concern to warrant more pointed consideration before I delve into the first case study, of the information developer Robert, in the next chapter. Both of these issues have significant ramifications for the day-to-day operations of the company, but, of particular importance to this study, each in its own way has a profound impact on the Aristotelian concepts of kairos and hexis for the information managers (and others) at Software Unlimited.

FACTORS AFFECTING KAIROS, HEXIS, AND THE ACHIEVEMENT OF PHRONESIS AT SOFTWARE UNLIMITED

While the full range of memory issues affecting the information developers' responses to situational exigencies (kairos) and development of long-term ethos (hexeis leading toward phronesis) will emerge only as we proceed through the analysis of memory work presented in the following chapters, a couple of these concerns require some background explanation that it will be helpful to introduce here. The first is the recent introduction of the Agile methodology, which has a pervasive effect on the ways in which virtually all of the employees of Software Unlimited must fulfill their job responsibilities—simply put, in the routine way they do things. The second concern is what I identify as a flaw in the chain of responsibility for user advocacy that appears very likely to be an unintended result of the company's prodigious growth over the two years prior to the study. I take both in turn.

THE AGILE METHODOLOGY, KAIROS, AND TECHNE

One particularly important recent development in the history of Software Unlimited to which the information developers frequently alluded was the adoption of the Agile software development process methodology. Software Unlimited had, like many traditional software companies, recognized the profit potential in the "cloud computing" model of software development and sales. In cloud computing, rather than purchasing "boxed" (i.e., sold as a disk in a box) software products on CD-ROMs in stores and then installing them on their own computers, customers instead pay for access to software as they need it and then access the software programs over the Internet, installing few, if any, portions of the program on their local computers. Responding to market forces and customer needs in a cloud computing model, not surprisingly, requires processes for developing new and upgrading existing software that are more nimble than those offered by traditional "waterfall" development processes in which long development and testing cycles are capped by code freezes to meet rigid shipping deadlines.

Upper management of Software Unlimited had recognized this need and had, over the year prior to my study, begun shifting its software development teams away from the slow waterfall model of software development to the quicker and more flexible Agile methodology. Management hoped that adopting the Agile methodology would enable more rapid upgrades and rollouts to respond to the cloud computing model. Implementation of Agile was ongoing at the time of the study, and different software product teams applied the principles of the methodology somewhat unevenly, as the following chapters will show.

Because the principles and the terminology associated with Agile software development recur throughout the remainder of the book, the methodology needs some explanation. According to Schwaber and Beedle's *Agile Software Development with Scrum* (2002), which articulates the version of the methodology employed by Software Unlimited (the company provided copies of the book to each employee and encouraged them to study it), the Agile methodology places a heavy emphasis on responsiveness to time. In this, the authors note, Agile is like the sport of rugby, from which the methodology draws its terminology: "Both are adaptive, quick, self-organizing, and have few rests" (Schwaber & Beedle, 2002, p. 1).

Further, according to Schwaber and Beedle, the Agile methodology borrows from process control theory the notions that all aspects of the software development process should be transparent to management and continuously empirically verifiable. That is, every employee's work on a project should every day be made visible in some fashion to every other employee concerned with the project. Two principal mechanisms enable this transparency: Scrum teams and Sprints. Scrum teams are small (ideally no more than nine persons), multidisciplinary, and, within limits, self-directed and self-organizing groups of workers who are tasked with "delivering new executable product functionality" during fixed thirty-day design periods, termed Sprints (Schwaber & Beedle, 2002, p. 9). The goal of each team, then, is to create a product or piece of a product that can theoretically stand on its own as an upgrade or stand-alone product during the thirty-day design Sprint. Each Scrum team must contain all the expertise necessary to complete these deliverables: "Regardless of the team composition, it is responsible for doing all of the analysis, design, coding, testing, and user documentation" (Schwaber & Beedle, 2002, p. 37). Consequently, every product team at Software Unlimited was composed of one information developer, one or two quality assurance (QA) team members, a Scrum team leader, a training specialist, and four to six software developers.

The thirty-day Sprints are bracketed by planning and review meetings in which team members, management, and users determine the functionality to be built by the team during the upcoming thirty-day period or review the results of the previous thirty-day period. The most important meetings and mechanisms for transparency prescribed by the version of the Agile meth-odology advocated by Schwaber and Beedle, however, are the Daily Scrums. The Daily Scrum is a short meeting lasting no more than fifteen minutes in which all team members meet face to face or via conference call and answer for each other three questions: "What have you done since the last Scrum? . . . What will you do between now and the next Scrum? . . . What got in your way of doing work?" (Schwaber & Beedle, 2002, p. 43).

The methodology asserts that the rules of the Daily Scrum must be ad-hered to strictly. These rules are that the location and time of the meeting be constant, that all team members arrive on time, that team members speak briefly and to the point when answering the three questions, and that any design issues or problem discussions be deferred until after the Scrum. The Daily Scrum, in other words, is meant to be a method for maintaining team and management awareness, not a working meeting. Although dif-ferent teams observed during the study adhered to the rules of Scrum with varying degrees of strictness, one factor that seemed common to all was the practice of holding short informal follow-up meetings immediately after the Daily Scrum in which impediments or design issues could be discussed by smaller subsets of team members. The information developers, in par-ticular, made frequent use of the Daily Scrums as opportunities to arrange such short face-to-face meetings with developers and other team members.

The Agile methodology, not surprisingly therefore, played a large part in determining the conditions of kairos for the information developers at Soft-ware Unlimited. Sprints required that information developers, like the rest of their teams, have something to show for their efforts every thirty days. Daily scrum meetings, especially the prohibition against being late, affected daily schedules and required that the information developers be prepared to make an account of their actions over the most recent working day. In short, the Daily Scrum placed a performative condition upon the technical skills of the information development team members: each information de-veloper was expected to have something to "show" or at least discuss issu-ing from the practice of his or her craft (techne) every day. Furthermore, the cross-functional composition of the Scrum teams meant that this work was subject to the ongoing scrutiny of team members with departmental goals, allegiances, and expectations different from their own.

USER ADVOCACY, HEXIS, AND PHRONESIS

For years, researchers have been arguing that technical communicators bring valuable insights about usability to product design processes and, therefore, should be included in product design decisions throughout the development cycle (e.g., Dilger, 2006; Salvo, 2001; R. R. Johnson, 1998). When I first began my research study on the premises of Software Unlimited, at first blush I became enamored of the notion that here, finally, at some point in this company's history, upper management appeared to have heard this cogent argument and, consequently, had, in an instance of foresight, made the task of acting as advocate for users a significant part of the job description of its information developers. And indeed, the information developers appeared to take this aspect of their jobs very seriously, as we will see later, and they all made repeated references to this part of their jobs throughout the study. In fact, several of the information developers emphasized that they believed user advocacy in product design to be a more important duty than writing documentation. The information developer Lucy even went so far as to say that "writing per se [is] the smallest part of our job." In short, because they believed user advocacy was one of their most important jobs, the information developers wanted knowledge about users of the company's products to be regarded as their domain of expertise.

However, despite the information developers' enthusiastic embrace of this aspect of their jobs and management's apparent support for it (implied by its inclusion in the company's actual job description), my research leads me to conclude that the apparent intentional inclusion of user research and user advocacy under the umbrella of responsibility of the information developers—at the particular stage in the company's evolution at the time of this study—is actually an accident attributable to rapid growth. It turns out that this situation is an artifact from the time when the company was much smaller and when, like so many technology start-up companies, the sparse, often undermanned staffs of such companies are forced (often quite happily) to wear many hats in what are often ragtag, loosely organized or structured entrepreneurial ventures. And in fact, not only was user advocacy not the sole the responsibility of the information developers, but user research in particular was not regarded as the information developers' province whatsoever. Usability testing was, instead, supposed to be conducted by a dedicated—and newly established—user-interface design (UX) team, and this team would be responsible for the dissemination of user testing data. I will have more to say about this peculiar situation in the case study chapter devoted to Angela (chap. 6), in which the difficulties presented by this flawed organizational arrangement become glaringly apparent, but suffice it to say for now that this discrepancy would play an important role in the aspirations toward achieving phronesis among the information developers.

THE CASE STUDY: INTRODUCING THE PARTICIPANTS

The case study presented in this book focuses on the five information managers or technical communicators who constitute the information development team at Software Unlimited. The team consists of four women and two men, each with varying degrees of experience in the field and different lengths of tenure with the company ranging from a few months to almost six years, giving each writer a different perspective on the memories of the organization. The team members chose or were assigned the pseudonyms Angela, Lucy, Monica, Peter, and Robert, and all of these individuals reported to Becky (also a pseudonym), the team manager, who will also be present as a significant factor in this research study.

Three of the information developers, Robert, Lucy, and Angela, participated in work session observations enabling close attention to situated practices. Sessions with these three information developers form the basis of my analysis and discussion in chapters 4–6. These particular participants were selected largely for pragmatic and circumstantial reasons, because they were the participants who appeared most enthusiastic about the study and most comfortable with being researched and, consequently, most consistently made themselves available for close observation of their activities.

However, the selection ultimately proved fortuitous, because the three team members' particular relationships to their communities of practice at Software Unlimited made them representative of the three stages of membership that an individual may hold within a community of practice: newcomer, journeyman, and master (Lave & Wenger, 1991). In addition, the situations that the three face are typical of situations faced by knowledge workers on many contemporary work teams. These relationships and situations include a newcomer trying to do a new job while also learning how to fit in, a midlevel professional coping with shifting responsibilities, and an old-timer in her communities of practices, attempting to do creative and cutting-edge work while mentoring junior members—and at the same time keep pace with a rapidly changing organizational landscape. We begin with the least-tenured information developer, Robert, in chapter 4.

CHAPTER FOUR

Learning Memory

A rhetorical attitude toward composition emphasizes copiousness, abundance, plenitude, aggregation, so that a composer will have something to say whenever an occasion arises.
—Sharon Crowley, 1993, p. 43

This chapter focuses on Robert, the most junior member of the information development team, who at the time of the study had been employed at Software Unlimited for just over two months. Because of Robert's status as a newcomer to the organization, this chapter offers an opportunity to focus on the challenges that newcomers face as they attempt to discover and use an organization's information—its memories—in order to participate in its activities and gain recognition as full members of its communities of practice. The exploration of these challenges in this chapter reveals that finding and effectively and appropriately using an organization's memories are not always easy tasks for newcomers, because newcomers face challenges of enculturation and socialization beyond the scope of detail of most new employee orientation and training sessions.

Unlike most of the members of the information development team, Robert does not hold a degree directly related to technical communication, nor has he worked full time in the job role of technical communicator before. However, he possesses considerable experience writing about technology (he has self-published two books about handheld technological devices), has some experience freelancing for technology companies, and possesses a master's degree in educational technology. These experiences, as will be seen, have made Robert comfortable with learning new technologies and have given him confidence in his ability to manage information effectively and efficiently. In fact, although the growth of academic programs is changing things, Robert's somewhat unorthodox career path remains fairly typical among technical communicators, many of whom "were either technically educated people who learned to communicate, or communicators who learned to deal with technology" (Davis, 2001, p. 140).

Regardless of their educational backgrounds or previous work experiences, to become a full participant in an organization, newcomers like Robert must absorb both functional and social knowledge of the communities of practice within their new organizations. That is, it is not enough for the newcomer to demonstrate technical competence (i.e., techne knowledge)—such as by showing excellent written communication skills or programming abilities—he or she must also demonstrate an "ability to read social situations and to act effectively within them" (Sullivan, Martin, & Anderson, 2003, p. 123). To read and write (i.e., to act) within these social situations, the newcomer must learn "local knowledge [including] the rituals and dynamics of practice, the tensions of institutional power and influence, and local rhetorical lore" (Sullivan, Martin, & Anderson, 2003, p. 127). These types of knowledge and their inculcation over time gradually move a newcomer beyond techne to the phronesis of a full member of an organization and the communities of practice to which he or she belongs.

Learning about the memory regime is a key aspect of this process. The memory regime governs access to local knowledge by determining the organization's approaches to, understandings of, policies regarding, and infrastructures for accessing its own collected information or knowledge. Learning the memory regime thus entails learning who possesses or has access to information, who is empowered to use information, in what arguments different kinds of information may be deployed, and when and under what conditions it is permissible to deviate from the organization's norms—all crucial components of the situated ethical awareness entailed in phronesis.

Both the memory practices the newcomer brings with him- or herself and the memory practices of the organization and the communities of practice within it play important parts in this socialization process. That is, rhetorical memory practices enable newcomers to discover information important to an organization, to preserve this information so that it can be refound, and to appropriately and effectively employ this information to bring value to the organization. Performing this work successfully and consistently over time demonstrates to others that the newcomer has become a full member of the organization. Rhetorical memory practices thus contribute to the transformation of the newcomer from novice to master: rhetorical memory practices prepare him to meet workplace exigencies with a confidence born of habit (hexis); to speak fluently (i.e., without "mushfaking" or stumbling through a discourse with which one is not comfortable or knowledgeable); to act appropriately at the right time without a script (kairos)—to, in sum, exhibit phronesis (Gee, 1990).

Despite Robert's short tenure with the company, at the start of the study the information development team appeared to have already begun to accept him as a full member of its community of practice. In particular, Robert's thorough research skills had earned him considerable respect with the other information developers, who praised his ability to bring value to the team. The information developer Peter, for instance, averred, "He's really great at research. We haven't worked together that long but he comes up with stuff. . . . I want to learn more about how he does research in the future because he seems to come up with a lot and come up with new ideas from places I never would have thought to look." Becky, the information development team manager, was even more emphatic in her praise of Robert's skills: "He's an unbelievable researcher. He's just like give him anything and he'll go to town. It seems to me that he really enjoys the challenge of learning new things and digging in and kind of finding out anything that he needs to know." In the opinions of his colleagues on the information development team, then, Robert appears well on his way to becoming a master of the organization's knowledge. In information theory terms, Robert appears to his information developer colleagues to be a consummate *puller* of information, an information omnivore always reaching out seeking new information to add to his stockpile and, by extension, to his "local" community of practice (Kirsh, 2000). Last, in rhetorical terms, Robert appears well on his way to achieving phronesis: his actions indicate that he is (and the other information developers perceive him to be) the type of creative, dynamic, and innovative individual the firm values most highly.

The information development team is only one of the communities of practice at Software Unlimited to which Robert must prove himself, however. Each of the information developers is assigned to be "documentation lead" on a cross-disciplinary product team. Adhering to the Agile development methodology, each product of Software Unlimited is developed by separate cross-functional teams made up of software developers, quality assurance engineers, user interface developers, a team manager, and a documentation lead—a technical communicator who is responsible for writing and coordinating all of the documentation for the product and product team throughout the design cycle. The documentation lead is also officially charged by the company to act as a "user advocate" with the other members of the development team.

Since his arrival at the company, Robert has been assigned as the documentation lead on Software Unlimited's newest product, a web-based version of an older conventional boxed (i.e., sold in a box at retail outlets) Software Unlimited product. The software developers, quality assurance engineers, user interface designer, and team manager in this community of practice are therefore the colleagues with whom Robert works most closely and for whom he must work hardest to prove himself capable of functioning as a full member of the organization.[1] However, because these teammates possess "different backgrounds, knowledge sets, and processes" and perform work tasks different from those of the information developers, they do not necessarily share the information development team's values and expectations about information and knowledge (Dicks, 2010, p. 60). As the observation sessions that follow demonstrate, in marked contrast to his reputation among the other information developers, Robert appears to remain, thus far at least, an unproven newcomer and an outsider to this community of practice.

Analyzing memory work centered on Robert, therefore, will help us better understand the ways in which newcomers adapt their personal memory practices to the strictures of a new memory regime and the communities of practice with which it is associated. As these observations reveal, Robert brings with him a rich set of memory practices and assumptions about research, knowledge, and information that inform his initial attitudes and approach to the memory regime or regimes of Software Unlimited. Yet, as he interacts with various teammates at Software Unlimited, Robert discovers that the organization and the various communities within it each have their own set of favored practices, some of which conflict with his own ideas. This ought to come as no surprise, given the fact that even companies that are predicated on the same model usually evolve their own unique ethos and memory regimes. And yet, the successes and failures of Robert's attempt to adapt to the specific memory regimes at his new company offer valuable lessons for understanding the difficulties that a newcomer faces when attempting to join an organization.

FIRST OBSERVATION SESSION: ARGUING FOR CHANGES TO THE USER PROFILES

When I arrive at 9:00 a.m. on the day we have set aside for job shadowing and observation, I find Robert already hard at work in his cubicle. Robert has his laptop open on his desk. This laptop is attached to an external monitor, keyboard, and mouse, and sits slightly to Robert's right while the exter-

nal monitor sits directly in front of him. He has enabled a function that allows his laptop display to act as a subsidiary display for the monitor so that
he can view information on both. Robert informs me that his main task for
the next hour or so is to start updating a section of the draft of the product
user guide to reflect changes in the user interface and workflow of his product. Robert further informs me that he is troubled by the latest version of
the user interface, which had been changed substantially from earlier iterations. As a result of these changes, he now believes the interface won't work
because it will be too complex for most users to update. He explains that
this is particularly frustrating to him because an important goal of his product team for the Sprint has been to make profile maintenance simpler and
more usable. He also adds that the "End of Sprint demo" of the latest version of the product to the whole company is scheduled for ten days hence,
so any changes to the user interface (UI) will have to be completed no later
than six days from now.

Robert's first action as he begins working on this task is to open the
most recently updated draft of the user guide and print a copy; however,
rather than working from this draft, he creates a new Word document and
starts from scratch. Robert informs me that he is starting anew because
the extent of the changes to the product have made him decide that it will
be easier to start a new document than to update the old, even though the
deadline is looming. He creates the section headings in this new document,
naming them "Set up users," "Create a new profile," and "Assign profiles."
Noticing that he is not referencing the printed version of his draft as he
writes these headings, I ask him if he is working from a memory of his
draft. Rather than working from memory, he informs me that he is instead
"thinking like an admin," or thinking about the prototypical administrator
tasks that a system administrator will need to perform to maintain user
profiles in the product.

As he begins to write the text for the "Create a new profile" section,
Robert extensively references the product interface, which is running on
his secondary laptop, but, again, he does not refer to his original draft or to
the printout very often. He works on this new document for about twenty
minutes, alternating his attention between the document in progress and
the product several times, and quickly writing about one page of documentation (see fig. 4.1) as he navigates through the product. At about this time,
Robert turns to me and informs me that he is beginning to perceive some
serious problems in the product workflow. He explains that these problems
may have occurred because there is no documented plan explaining how the
interface should work: "The UI designers just decided to create it." At this

Create a New Profile

To create one or more profiles, log in to the ▓▓▓▓▓ ▓▓▓ Administration Site.

1. Click the **Profiles** tab.
2. Click the **Add** button.
3. Enter a Title in the Title field

 The title of the profile will be seen by the Presenter in the ▓▓▓▓▓ ▓▓▓ ▓▓▓▓. The title should be descriptive and make sense from the Presenter's point of view. For example "Profile 1" might not be as useful to the Presenter as, "Podcast to LMS."

4. Optional: Enter a description of the Profile.
5. Check the Default Box if you know this Profile will be...

Determine Record, Encode, and Publish Settings

The default recording setting is 5 frames per second (FPS) and includes audio recording.

1. To change the record settings, click the change link. A list of available options is displayed.
2. Click desired record setting from the drop down menu and click Save.
3. If a desired option is not listed, a new record setting must be created. See [link to this topic].

 A frame rate of 3-5 frames per second is often suitable for typical PowerPoint or Keynote presentations. Higher frame rates are better suited for presentations with more motion on-screen. Higher frame rates typically result in larger file sizes.

4. When satisfied with the record settings, click the Add Encode Settings Link. A list of encoding options appears.
5. Check the boxes next to desired encode settings in the list and click Save. There is no limit to the number of ways a recording may be encoded.
6. If a desired option is not listed, a new encode setting must be created. See [link to this topic].
7. Click Save.
8. Click Add Publish Settings for each encode setting. A list of available publishing destinations is displayed.
9. Check the boxes next to desired publish settings in the list and click Save. There is no limit to the number of publishing destinations per encoding.
10. If a desired option is not listed, a new publishing destination must be created. See [link to this topic].
11. Click Save.
12. Click Save.

Figure 4.1　New section of the user guide—"Create a new profile" procedure.

point, he stops writing and, with an intent look on his face, begins exclusively navigating and experimenting with the product interface.

Robert next informs me that he has already had a conversation with one software developer about the workflow for this procedure but that he cannot remember exactly what she had told him about it, so he is not entirely sure if the software developers are planning to change the interface again or not. So he creates another Word document and writes down a particular question about the workflow for the developer, informing me that he will ask her this question after the day's Scrum meeting. After writing his question, he resumes documenting the procedure and stepping through the product interface. However, after about one minute of work, he realizes that he has another question for the software developer and then another and another and another. He updates his second document (which I hereafter refer to as his "reminder document") with these questions as he works, switching between the product interface and the two documents multiple times. After about thirty minutes of this, he opens a screen capture program and begins taking screenshots from the user interface to clarify his questions in the reminder document and adding annotations and callouts to these screenshots as he deems necessary (see fig. 4.2).

While working, Robert stops briefly to create a (handwritten) Post-it note to remind himself to ask his colleague Peter to show him the documentation for Peter's product. As he does this, Robert informs me that he likes the user interface of Peter's product, which he believes "makes the doc almost unnecessary," and would like to see a similar solution for his own product and documentation. He further adds that, based on the problems with the workflow and the number of questions he has for the developer, he suspects that the procedure he is documenting for the user manual is going to change substantially.

Robert continues working in this way for the next half hour until an Outlook calendar reminder pops up on his primary screen, giving him five minutes' notice that his Daily Scrum is about to start. He prints his reminder document and then saves it to his Windows desktop, the location he had noted during our initial interview in which he tends to save documents that have only a limited lifespan. Next, he saves the new draft of the user guide to the network drive. He then grabs a pen and his printout of his reminder document and we hurry off to the Daily Scrum meeting so that we will not be late (the rules of Scrum discourage tardiness). As we hurry to the Scrum, Robert informs me that he uses Outlook reminders because he has trouble remembering to attend his Scrums now that they are held on the other side of the building rather than directly adjacent to his cubicle. He further states that the Scrum has been "a motivating factor this morning."

As the members of the team gather in a semicircle around a large Gantt chart in the "Scrum Hall," several of them, including Robert, review and update their statuses (see fig. 4.3). They do this by sticking new Post-its on a

Scrum questions for ▮▮▮

Is there a limit to the title and do you know if the recorder plays fair with it?

Should you be able to modify a default profile? If you can is it still default? They could screw it up, and maybe not get back to the default. Also, how do we make default publish settings? (except maybe in the case of a trial where maybe we can assign them a ▮▮▮▮▮ acct)

One challenge is that the profile names try to be descriptive in both functionality and who they're for. Talk about details in the recorder.

Who sees the Profile description, I am assuming only the Administrator.

If I check the Default box, how does that impact assigning profiles to users if at all?

Are default Record, Encode, and Publish settings really just called "Default"? How do I learn what default means?

It's really cool you can click the grey "Add Encode Settings" link. Wondering if we should get rid of this part, and just give them one option? Maybe move the link up, left?

Figure 4.2   Robert's reminder document containing questions to ask the software developers after the Daily Scrum meeting.

large, communal whiteboard. I take up a spot somewhat behind and outside the semicircle of team members that is forming facing the chart.

The team leader starts the meeting promptly and makes a few general comments about the team's status. Robert takes this opportunity to introduce me to the entire team before the team members, one by one, begin to make their three-part report about what they had worked on the previous day, what they are planning to do today, and whether they are experiencing any impediments to this work. The rules of the Agile methodology, which dictate that the Daily Scrum not become a working meeting, prevent Rob-

Figure 4.3 Scrum Hall. The Agile methodology encourages Daily Scrums to remain short, so a hallway with no chairs can be a desirable location.

ert from addressing his issues with the particular developer involved on the project in the Scrum itself, so, when it is his turn to speak, he simply notes that he believes the current iteration of the workflow is problematic (i.e., that it is an impediment to his work of writing the documentation) and asks the developer (Andie) if she has time to meet with him afterward. Andie informs him that she does have time, so after the Scrum, Andie and another software developer, Gene, remain behind to talk with Robert. The three gather in a small huddle and I take up position slightly behind them. Having heard Robert's concern in the Scrum, Andie and Gene begin by reassuring him that the version of the product interface that he has been struggling to document will indeed be changing again. At this point, Andie asks to see Robert's reminder document and looks it over for about a minute before, using her notebook as a writing surface, with Robert's pen, she begins writing answers to several of Robert's questions and sketching a small mockup of the new interface in the white space at the bottom of the first page (see fig. 4.4).

As Andie reviews Robert's document, she and Gene explain how the new version of the interface will function and why they believe this new version will resolve the issues Robert has identified. Robert listens to their explanation before noting that the workflow as they articulate it will still be confusing to users. Andie responds to this assertion by asking Robert if he can give them any advice. Robert articulates his vision for the interface and there is some back-and-forth discussion of the workflow, with considerable attention paid to Robert's screen captures and Andie's sketch, and much gesturing as all three participants gesture to trace competing potential interfaces in the air for each other. Robert, perhaps because he is no longer holding his reminder document or a pen, gestures frequently as he attempts to articulate and argue for his version of a redesigned interface. After a few minutes of this, all three fall silent for a while as Andie and Gene look at the page and appear to be thinking through the implications of what Robert has told them.

Finally, the developers admit that they agree with Robert that there might be problems with the new version of the workflow and that they will try to correct these problems. To address the problems, Andie draws another sketch in the space at the bottom of the second page and shows it to Robert. Robert looks at the sketch for a few seconds before informing the developers that he believes it represents a good solution to the usability problems he has identified. Robert leaves the page of the reminder document that contains the sketch with Andie so that she and Gene can use it as a guide to making the changes to the interface. On our way back to his cubicle, Robert confides to me that he is pleased at the way the meeting has turned out.

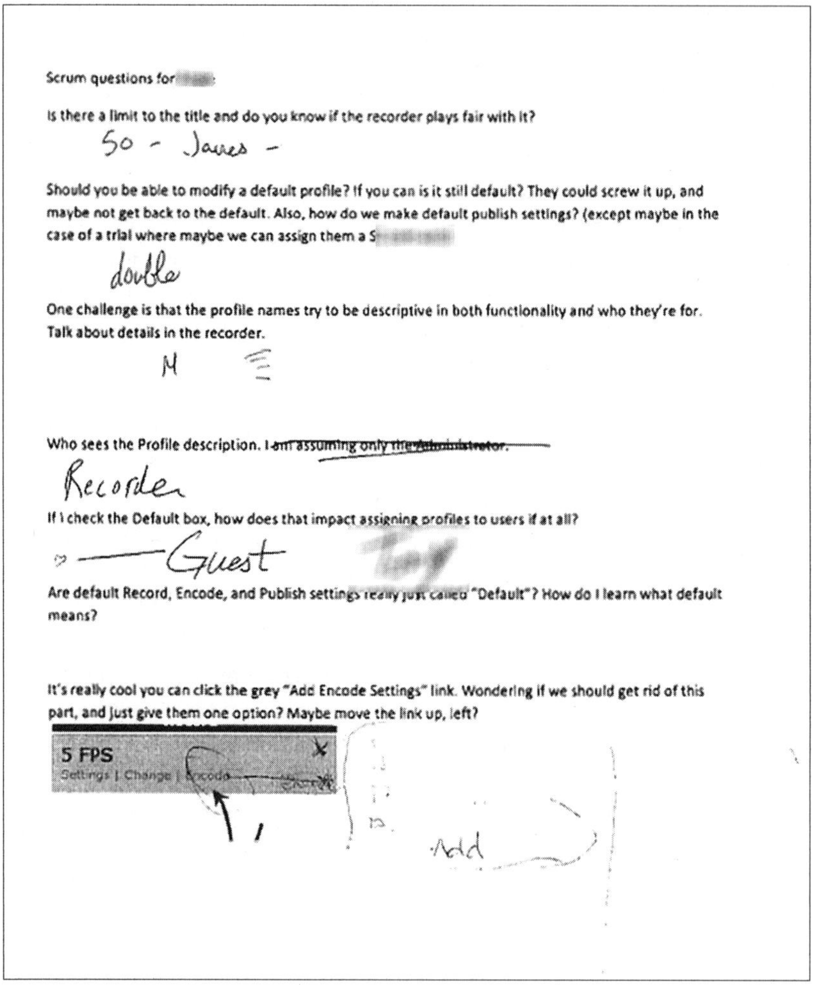

Scrum questions for [illegible]:

Is there a limit to the title and do you know if the recorder plays fair with it?

50 - James -

Should you be able to modify a default profile? If you can is it still default? They could screw it up, and maybe not get back to the default. Also, how do we make default publish settings? (except maybe in the case of a trial where maybe we can assign them a S[illegible]

double

One challenge is that the profile names try to be descriptive in both functionality and who they're for. Talk about details in the recorder.

M

Who sees the Profile description. ~~I am assuming only the Administrator.~~

Recorder

If I check the Default box, how does that impact assigning profiles to users if at all?

Guest

Are default Record, Encode, and Publish settings really just called "Default"? How do I learn what default means?

It's really cool you can click the grey "Add Encode Settings" link. Wondering if we should get rid of this part, and just give them one option? Maybe move the link up, left?

5 FPS
Settings | Change | Encode

Add

Figure 4.4 Annotations to Robert's reminder document. During an impromptu meeting in the Scrum Hall, Andie and Gene write answers to Robert's questions in the margins.

SECOND OBSERVATION SESSION: MEETING WITH WALLACE

About one month after the first observation session, Robert again invites me to observe his morning work session and subsequent team meeting. When I arrive at 9:00 a.m., Robert explains to me the purpose of the activities he will be engaged in during my observation. As in the first session, he is concerned with a user interface and workflow problem that he perceives in his product. In this case, the issue concerns the manner in which this product, a web-based product, will be integrated with the much more complex and full-functioned boxed version if users attempt to use both versions on a single computer. He informs me that he has been made aware of the issue by one of the customers currently involved in the beta testing of the product. This beta user has posted a message to the beta wiki, which Robert moderates, indicating that she is confused by several aspects of the integrated interface. After reading this user's post, Robert has grown increasingly alarmed that the integration of the two products is not being thought through with enough care by management or his teammates—a notion reinforced by a meeting of product managers that he attended the previous Friday. In this meeting, management appeared to believe that integrating the two versions of the product would be "an uncomplicated affair."

Robert's purpose during today's work session, therefore, is to update a PowerPoint presentation to take to a meeting with Wallace, "a talented web developer," on his team after the Daily Scrum. Robert hopes to enlist Wallace as an ally in his effort to change the integration interface. Robert believes that Wallace will have particular clout because he is the person in charge of all the interactions with beta testers for their product. However, Robert acknowledges that winning Wallace to his cause might be difficult because Wallace "gets pulled in two directions a lot."

Robert begins his work session by opening a PowerPoint presentation that he had created on a previous day. In our follow-up interview, Robert will inform me that he had created and delivered this first portion of the presentation to his teammates on the information development team; the team manager, Becky; and their director the day before so that they would all be "on the same page" when talking to the rest of the company about the issue. Even though his meeting with Wallace will be a one-on-one dialogue instead of a presentation delivered to a group, Robert opts to update this PowerPoint instead of creating a new document from scratch. Today, Robert plans to add slides containing sequential screenshots depicting step-by-step walk-throughs of two scenarios users might encounter when attempting to use the integrated product interface.

Robert next opens his screen capture software and the product itself and proceeds to navigate through the product's interface, taking screenshots as he goes and pasting these screenshots into the presentation. For speed, Robert begins work by pasting every screenshot onto a single slide, with the intention of moving each image to a separate slide when he finishes screen capturing. Soon however, Robert informs me that he is afraid he will forget the proper sequence, so he begins alternating between the product interface and the presentation, cutting and pasting screenshots one at a time into

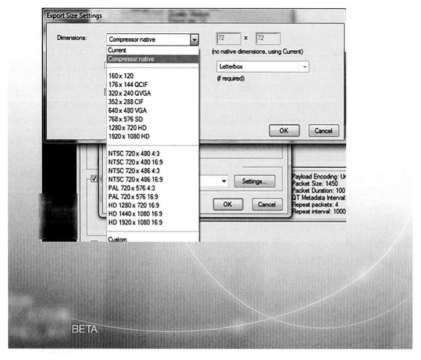

Figure 4.5   A slide from Robert's PowerPoint presentation displaying
one step of the product workflow sequence.

separate slides and sequencing these slides to imitate the functioning interface for the scenario he wishes to show to Wallace (see fig. 4.5).

When Robert finishes this, he polishes up his slides by adding explanatory text and graphics to enhance the visual appeal. Finally, he informs me
"that's good enough for Wallace" and prints it. As before, Robert has worked
right up until his Outlook reminder informs him that his Daily Scrum is
about to begin. He grabs a pen and the printed version of the PowerPoint,
and we leave for the Scrum Hall. As in the first scene, the Scrum meeting
itself is short and Robert uses it mainly as an opportunity to remind Wallace
that they are planning to meet afterward.

Immediately after the Scrum, Robert, Wallace, and I walk to Wallace's
office, which is nearby in the software developers' wing of the building.

When we arrive, Wallace begins by asking the purpose of the meeting. Robert recaps for him the previous Friday's meeting in which the product integration was described as "seamless" with, Robert believes, no consideration of the development effort involved in making it so. Specifically, Robert says the overall opinion of the managers came across as "Make it happen, developers. Make a button or something." At this point, Robert places his PowerPoint printout on the desk so that it faces Wallace. Over the next few minutes, Robert steps Wallace through the workflow in detail, pointing frequently to the screenshots with his pen and alternating his attention between the printout itself and Wallace.

After Robert finishes articulating his concerns during this walk-through, Wallace counters by asserting that he is not particularly worried about the issue Robert identifies. Wallace believes that the issue is not a new or unique problem, and he calmly makes this point by invoking his superior knowledge of the product's history: "It's not necessarily a new problem. . . . To me this is not really a new thing; this is the same kind of drawback [as in the previous versions of the product]." Wallace also utilizes Robert's PowerPoint printout to make his points, gesturing to different slides at different points. Finally, Wallace asserts that he considers the problem more a matter of providing adequate documentation than of correcting the workflow in the software itself.

Robert holds his ground, however, countering each of Wallace's points by appealing to his own knowledge of user preferences. At several points during this counterargument, Robert speaks in the person of the user, saying things like, "If I'm using [the product] and I want something and then I do all that work and I send it out there and I don't understand if I get surprises then I'm going to get so mad because the profile took out something or [the user will think] 'Why is it like this?'" Robert also attempts to convey to Wallace his own vision of what an improved interface and workflow would look like by alternately pointing to his screenshots and tracing pictures of a revised interface in the air using gestures.

Finally, after about twenty minutes of similar back-and-forth discussion, Wallace half-heartedly concedes that Robert may have a point: "I can see the benefit. Now, I don't think that they [users] are going to do this a lot, but I could see the benefit." The meeting concludes, leaving Wallace, at best, a reluctant ally. In our follow-up interview, Robert informs me that he will have to continue his efforts to gain traction on the issue with others on his product team: "So I'm going to fix this [the presentation] up and try to better define that [workflow problem]."

ROBERT'S MEMORY WORK

In these two detail-laden sessions, we observe Robert engaging in a variety of activities employing a rather broad range of workplace affordances, and doing so in a manner designed very purposefully to help him write about and contribute to the design of his team's software product:

Robert attempts to "think like an admin" (i.e., a system administrator) in order to remember topical subheadings when rewriting the user guide, an example of a memory schema for *finding* or "inventing" content.

As he writes, Robert references the software product, which he displays on his second computer monitor, *reminding* himself what he is writing about.

Robert creates handwritten Post-it *reminding* notes, in a limited but strategically targeted way, so that he will not forget the questions he needs to ask the software developers after the Daily Scrum, to ask an information development team colleague a question, and last, to update his status on the board in the Scrum Hall.

Robert quite proactively convenes, or perhaps more accurately, requests (and receives) meeting opportunities for *referencing* the knowledge of other team members, to which he brings an arsenal of documentation in support of the position he intends to take within those meetings.

Robert painstakingly recreates a mockup of his software product's user interface in PowerPoint to serve as a script during his meeting with Wallace, *reminding* him of his talking points.

Like any employee working for a fast-paced, growth-oriented company, Robert must engage in these practices against the backdrop of infrastructures, policies, and conventions that, together, constitute the overall memory regime of Software Unlimited, which, as I have described it in chapter 3, consists of a more complicated set of subculture memory regimes, overlapping but independent, and falling loosely along the lines of the various communities of practice that coexist within the organization. However, what is perhaps most remarkable about Robert, which can be seen from just a preliminary review of these two observation sessions, consists of his poise, extreme confidence, and bold forthrightness in pursuing his agenda and particularly in his amicable dealings with senior members of his team— remarkable for someone just two months on the job. It may be somewhat preliminary to say at this point, but Robert seems in the early going to be able to effectively cut through any political cobwebs or entanglements to focus strictly on the real business of Software Unlimited: improving and expanding its product line. As a result, these first two observation sessions with Robert, which we consider first in this book, may be the optimal ones to consider in our effort to begin describing, with some specificity, some of the characteristics of this complicated galaxy of merging yet at times competing memory regimes, specifically, by attempting to answer the first two and most fundamental research questions that motivate this study:

> What types of organizational information are the most important for Robert and the other technical communicators? and
> Where does this information reside, and, by extension, how does it move through the organization?

To answer these questions, we can turn to the evaluative framework that I outlined in chapters 2 and 3. Rhetorical theory identifies two categories of practical or situated knowledge: techne knowledge, or knowledge about how to do or build something, and phronetic knowledge, knowledge about how to be or act in social situations. The principal measure of mastery in a techne lies in a person's ability to respond to kairos by doing the right things or taking the right action at the right time, while the—rather more difficult to observe—measure of phronesis lies in a habitual disposition to behave ethically and responsibly in social situations.

At Software Unlimited, one of the principle determiners of kairos results from the time pressures arising from the Agile software development methodology: product teams must meet their thirty-day Sprint goals, and product team members must be prepared to discuss their status and impediments in the Daily Scrum meetings. The two sessions above reveal multiple instances of Robert responding to kairos arising from the Agile methodology (e.g., Robert hurriedly prints his list of questions to take to the developer meeting; he fears that the rush to meet Sprint deadlines has caused management to overlook a major usability issue). Further, they demonstrate Robert responding well to these exigencies (e.g., Robert writes quickly and well, enabling him to use documentation authoring as a time for diagnosing potential usability problems in the software interface). In other words, it appears that Robert has arrived at Software Unlimited with a pretty firm grasp of the techne knowledge necessary to succeed as a technical communicator.

At Software Unlimited, concerns of phronesis for the information development team appear to be most visibly influenced by the conflict created by the firm's mandate to its information developers to act as both documentation writers and user advocates on their software design teams. I will discuss this issue in more detail later, but, in brief, the information developers, particularly newer information developers like Robert, are almost entirely dependent on the goodwill of the software engineers on their teams for all-important information about their product. Yet the mandate to advocate for user-centered changes places the information developers in a potentially adversarial relationship with their benefactors: changes to user interfaces could lead to bruised egos, missed deadlines, and extra work for the software developers. These conflicts, which principally arise at the interfaces between the communities of practice to which the information developers belong, present knotty ethical conundrums affecting phronesis. Which professional identity and ethos should the information developers cultivate: technical communicator or user advocate?

From the data in these sessions and from our interviews, we are able to make the following general statements about Robert's memory practices: (1) Robert believes that careful, even meticulous archiving practices will make him successful in both the short and the long term in his organization. (2) Although largely invisible to Robert, reminding practices rather than archiving practices are essential in helping him function in his fast-paced work environment. The influence of the memory regime—and here I am arguing that the sub-memory-regime of the information developers that is specifically in play—can be seen in these behaviors. Viewing these practices through the lenses of rhetorical theory makes the functioning of this subculture regime explicit. In the following sections, I will explain how an analysis of the data collected from the sessions of memory work, as well as from the interviews with Robert and the other information developers, leads to these conclusions.

ROBERT BELIEVES METICULOUS ARCHIVING PRACTICES ARE THE KEY TO SUCCESS AT SOFTWARE UNLIMITED

Given Robert's newcomer status, the fast-paced work environment encouraged by the Agile methodology, and the complexity of his relationships with his colleagues on the software development team, perhaps the most striking aspect of Robert's performance in the scenes recounted above, as I have already noted, is the degree of poise and self-confidence he displays. Despite his newcomer status, Robert believes that he has already learned enough about his software product to sway his much more senior teammates into making changes to its design. This sense of agency appears to be the result of several related factors: Of course, it is fairly clear that Robert's "base-line" confidence stems from his prior education and experience, and the fact that he is a published author in this field. But I maintain that Robert's prior experience merely "sets the table" for his forthrightness and demonstrated self-confidence in dealing with his colleagues, and that there are two other factors that are more pertinent to the research reported here. The first is Software Unlimited's assertion that it offers every team member, no matter how new, equal access to information about the company's evolving software products and an equal voice in making decisions about those products; and the second is the faith Robert places in his own research abilities and memory practices. For this latter reason, we need only reflect on the diligence with which Robert conducts his research efforts through finding practices, creates archives and reminders, and ultimately uses these efforts to persuade colleagues to adopt his point of view.

Of primary importance is the spirit and the actively pursued and promoted culture of Software Unlimited. On its website and in its hiring literature, Software Unlimited portrays itself as a progressive company that encourages all of its employees to innovate and to contribute their insights to making better products and helping customers. All levels of management, including Becky, the information development team manager, echoed this aspect of the organizational culture. Becky stated that innovation is "the spirit of [Software Unlimited] in general—it's the culture that you are encouraged to definitely try new software and do different things." Robert echoed this confidence in the egalitarian spirit of Software Unlimited: "I think it's a lot more intellectually stimulating than the past [i.e., in other companies he has worked for], so there's a lot of good issues. And everybody is right sometimes, so it's really intriguing."

Because he believes that the company encourages and rewards innovative thinking, Robert has adopted a set of thorough archiving practices that he believes will not only aid him in his current job but will enable him to make a difference in the future (e.g., to move beyond techne to phronesis). In pride of place among these archiving practices is Robert's "back of the mind." Since he first arrived at the company Robert has maintained a folder on his laptop hard drive titled "back of the mind." "Back of the mind" is the archive where he stores documents containing information or ideas that he regards as potentially useful but for which he has no immediate or clear use in his current work tasks: "[It's] one thing I do that's been really valuable—I don't know if other people do it. So I have this folder called 'back of the mind' where I write things like crazy ideas or just things that I think would be cool to do some day but right now they'd be crazy."

Robert notes that, although he believes himself to be a natural work-aholic, someone who is always thinking about work and brainstorming ways to do his job better—or, more specifically, envisioning or entertaining novel, free-thinking innovations that might make his company's products better—he had never before bothered to maintain any similar archive of ideas before coming to Software Unlimited, largely because he did not believe that any of his prior employers would have valued those ideas. He immediately perceived Software Unlimited to be different: "Here you could actually do something. It's conceivable that some of those crazy ideas—that I might get a chance to do some of them. So it's worth writing them down, whereas at some companies, it's like don't even bother thinking about them or anything that's outside of what's the norm."

Thus, it certainly appears that Software Unlimited's very deliberate and highly focused corporate effort to establish its specific identity and mission, what in another context might be defined as establishing its brand, has been quite successful, at least insofar as attracting individuals who want to be a part of that vision. But what is most important for our purposes here is that Robert immediately created his "back of the mind"—a distinctly and objectively identifiable archiving memory practice—because he believed from the start that Software Unlimited fosters a culture in which employees are not just permitted but actively encouraged to contribute to ongoing design decisions and innovations. The organization seeks to communicate this aspect of its memory regime in diverse ways: from the infrastructures that the company provides for saving documents, to which all employees are given access, to its employment contracts and recruitment literature, which include language encouraging employees to think strategically about the organization. Accepting these assertions, Robert believes that the "back of the mind" will be a key strategic resource in his effort to establish his career at Software Unlimited. As Robert notes, this is an idiosyncratic practice: the data retention policies of Software Unlimited do not require him to maintain this information, and, to the best of his knowledge, he is the only one among his colleagues who does so.

There is something else worth noting here. Robert's creation of his "back of the mind" archive is also motivated by the fact that Software Unlimited, as it turns out, has no formal, company-wide, *centralized* content management system for collecting, backing up, or archiving the innovative ideas that are created or conceptualized by individual employees on their own. While the firm's invitation in its hiring literature—indeed, its challenge to employees—to be creative and innovative is laudable and a factor that appears to attract bright, talented people like Robert to join it, the company's free-form approach to innovation and creative development means that, in practice, the collection or "capture" of that innovation and creativity is left largely to individuals, and shared, if at all, mostly among the members of the various product teams.

This analysis supports my contention in chapter 3 that Software Unlimited is a company that, for better or for worse, does not have a strict, hierarchical, top-down and centralized memory regime emanating from upper management and flowing evenly and symmetrically down through the ranks of increasingly lower-echelon employees. Instead, there exists within the company a set of overlapping memory regimes whose boundaries are largely defined by the different communities of practice within the organization. As a result, the innovative and potentially valuable information that we have been discussing here tends to reside at the subcultural level of these various communities of practice, where some of it spills over into other communities by way of this "overlapping"—but not necessarily all of it.

Two other facts from Robert's observation sessions support this assessment. First, Robert's "back of the mind" is not the only individual, and thus discrete, archiving memory practice that we will see as potentially representing a direct if unconscious response or reaction to the lack of a centralized memory regime at Software Unlimited. As we will see in chapter 5, Robert's coworker and fellow information developer Lucy keeps a "hard copy" handwritten, paperback notebook in which she records in great detail the historical information—the "reasonings" as she calls them—underlying the present evolutionary iteration of her product lines. Like Robert, Lucy believes that keeping her notebook makes her a more valuable member of the community, initially by putting at her fingertips a wealth of historical knowledge that *other team members do not have access to*, but, most importantly and again very much like Robert, Lucy, through her notebook, acquires the authoritative confidence to contribute more substantively and vocally in team meetings. As Lucy states, "I know that I know what's going on. So I speak up much more often than I used to." Lucy, unlike the others, knows the "known unknowns." Furthermore, even if Lucy does not specifically acknowledge that this is the case, it is reasonable to conclude that, for all of its proactive and instantaneous "knowledge-at-the-fingertips" utility, Lucy's creation and keeping of her notebook is to some significant degree a direct reaction to the lack of any centralized, company-wide, archive.

The second fact supporting my assessment can be briefly stated. Specifically, while the Agile methodology was implemented as a means of ensuring that information was shared within and across departments (via the Daily Scrums and the End of Sprint demos), it was not designed nor intended to

serve as a mechanism for sharing the kinds of insights that Robert preserves
in his "back of the mind" archive or that Lucy preserves in her handwritten
notebook. As I have said elsewhere, I do not believe that these actions, or
more accurately, inactions, are in any way deliberate or sinister attempts to
hoard valuable information for their exclusive benefit, such as, for example,
to be used a means of ascending the ranks of the company. And in fact, we
have seen in this chapter that Robert readily shares his "back of the mind"
ideas with his teammates when it is appropriate, and we will see in chapter
5 that Lucy does the same, sharing her notebook information willingly.

However, neither Robert nor Lucy seems to feel required or even compelled to take the extra step to commit the discrete, individualized information and knowledge they alone possess to a public database or system
so that it can be accessed and viewed by every other individual in the company, either team members or non-team-members. And indeed, there are
no company or Agile rules or standards of operation that require them to
make this information available to everyone. Further, as we will see when
we discuss Lucy's case in chapter 5, there are no company-instituted auto-
mated memory or backup systems in place through which Software Unlimited could ensure that important company-wide knowledge is routinely
preserved in any sort of open-access, fail-safe way. That both Robert and
Lucy are not motivated to share their individual information outside their
immediate community of practice is a further strong indication that they
perceive the memory regimes that are of highest importance to them, and to
their job performance, to be those that pertain to their particular communities of practice—and not any sort of overarching corporate memory regime.

Still, this issue of multiple memory regimes aside, Robert appears to believe that the "back of the mind" will prove valuable to him in establishing
a reputation as a master of the company's knowledge—both within his communities of practice and with the company as a whole. We might, in short,
consider the "back of the mind" a component of phronesis knowledge, the
kind of knowledge accreted over the long term and existing for, essentially,
ethos- and identity-building purposes.

As for the contents of the "back of the mind," Robert identified diverse
criteria for inclusion. First are documents that he believes might serve an
agonistic purpose, possibly by helping him warrant arguments about usability with his software team: "If there's a big debate—like the one over
localization, then for that I'll actually take the time to write up." Of course,
there is a pragmatic aspect to this as well. Knowledge enhances persuasion
by putting the knower in a position of strength and perceived authority.
Second, Robert preserves "things that nag at me a lot," such as unanswered

questions or unresolved disputes. Third are documents that Robert hopes to eventually share in order to grow the knowledge base of the company, like the document "Lessons learned the hard way," which contained a list of "the top ten things that were difficult for me to do . . . when I first came here."

These last two categories support my assessment that there is no malicious or aggrandizing intent in Robert's creation and maintenance of his "back of the mind" file; in fact, by contrast they support the notion that Robert is actively inculcating a phronesis appropriate to the culture of Software Unlimited: saving this knowledge will, Robert believes, enable him to help make the firm better and assist his colleagues in their shared enterprise. That is, the second category of things that "nag" Robert may in fact turn out either not to be problems, or to include some of those kinds of issues that all employees eventually encounter in their working lives, things which never seem to admit of an acceptable resolution and so simply get worked around. Thus, until such time as Robert might have an epiphany that offers a brilliant solution to one of those "nags," there probably would be no benefit to making them public. The third category represents formulations of ideas that Robert does consciously hope to share with the company to try to help other employees and team members, and to help the organization not just to grow its knowledge base, as I've stated, but also to improve its overall operating efficiency—for example, in identifying and collecting experiential knowledge that may facilitate a more seamless integration of new employees into the memory practices of the organization, or of its subcultures. But these issues, too, are in Robert's estimation "not ready for publication."

We may presume that, at the appropriate time, and once Robert has an opportunity to organize and codify these ideas, he may then proceed to share them with his colleagues, or perhaps take them directly to management for formal consideration. However, as a final point, it should also be clear that at least some of these issues technically fall outside the realm of Robert's primary job description, and as we have seen in the observation sessions with Robert (and will similarly see later in the observation sessions with Lucy and Angela), Robert has his hands full just keeping up with the pace of the workload at Software Unlimited as it is. At a minimum, however, and taken together, the various types of document that Robert stores in the "back of the mind" will, he believes, enable him to be (and to appear to be) proactive and innovative during his career at the company. Their ends signal them to be, in other words, phronetic knowledge.

The "back of the mind" is not the only archiving practice that Robert engages in, however: he also maintains a separate directory for each release of his product. In keeping with the classification system of the organization, Robert labels these directories by Sprint number. Within these directories are subfolders for product deliverables (e.g., help files and .doc files containing user guides) and a folder that Robert titles "supporting materials." Robert maintains these Sprint directories principally for tactical reasons to meet the demands of kairos as he completes his daily work activities. This archiving practice contributes, therefore, principally to Robert's techne.

Robert does not appear to regard the tasks of writing the documentation deliverables or maintaining these deliverables in the product archives as particularly difficult or intellectually challenging: "It's not a big deal. I think like with the core deliverables . . . those deliverables, those go in a very special place—like it's all perfect and that's no problem." In other words, and as I have already noted, Robert believes himself to be in firm command of the techne knowledge of the technical writing aspects of his job.

But each Sprint generates a lot of other miscellaneous documents, which, although they are not deliverables, Robert considers to have potential value for meeting a variety of exigencies associated with the specific release. Figuring out ways to archive these supporting materials so that they can be found and used to meet future exigencies appears to be more challenging to Robert. Consequently, within each Sprint folder, Robert creates a subfolder titled "supporting materials," which serves as a general-purpose catchall for miscellaneous product and release-related documents (e.g., outlines, project plans, and meeting notes) that may contain important background or source material.

Alone, the practice of creating a catchall container for unclassifiable information is not unusual, but Robert is always thinking strategically, always trying to anticipate future informational needs. Therefore, Robert regards his "supporting materials" folders as a sort of strategic reserve that will enable him to appear to his colleagues, particularly his colleagues on his product team, to have mastered the knowledge of his product: "Stuff that goes in here is stuff that might someday be useful or it's some note that I had. Like sometimes people ask me if you have anything for a session description or for a flyer. Anyway, this is just three sentences or two different sentences that have pros and cons so that in one breath of air you can explain it."

Robert's archiving practices are, in fact, both tactically and strategically motivated. Tactically, his archive of information serves as an important resource to help him meet the day-to-day demands of composing and arguing with his teammates. Strategically, Robert believes that his archiving practices, particularly his "back of the mind," will eventually enable him to take on a leadership role in the organization. Together, his various archiving practices help Robert meet the exigencies of kairos while also cultivating his identity as a master of the organization's information—his pursuit, going forward, of phronesis.

Curiously, however, to almost the same degree that Robert emphasizes the importance of his careful archiving practices, he downplays the importance of reminding practices in helping him become an effective employee.[2] In Robert's own account, reminding plays only a minor role in helping him manage and use information. For instance, when questioned about the reminding tools offered by Outlook, such as the calendar function (which he utilizes in both observation sessions recounted above) he says: "I know some people use Outlook way more efficiently than I do. I only use email. . . . I don't use the other stuff." And, even though he is willing to concede that the Outlook calendar is "better than nothing" and that others might use it, he did not see it as a very useful memory tool: "It's really funny how everyone is so chained to their Outlook. I guess you have to be and one good thing is that if you have a meeting scheduled to end at two then it probably has to end at two because they have a meeting somewhere else. I guess it keeps things from dragging on a bit."

Robert similarly dismisses another ubiquitous tool for reminding in contemporary workplaces, Post-it notes, by emphatically asserting twice during our initial interview that, although he uses them for personal purposes, he does not use them very much at work. (Interestingly, however, even the brief slice of Robert's working life captured in the scenes above show him creating Post-it note reminders.) Instead of ephemera like Post-it notes, Robert indicates that, for work-related purposes, he prefers rolling his reminders into longer-term solutions (i.e., archiving practices): "Like the matter of that Skip button, it consumed me and so what will happen is I'll just be at home, or in the morning on the drive in, or in the shower or something like that, and I'll have, like, this list of things in my mind I don't really write down on Post-it notes or scrap paper that much and then I come here and I'll write it [on his computer] and I'll stick it in this folder [i.e., the "back of the mind"]."

As I questioned Robert about this preference for archiving over reminding, his answers indicated that, in his view, reminders lack two characteristics that he values: classification and contextualization. The first, classification and its attendant phenomenon of sequencing (see Bowker & Star, 1999), relates to the permanence, stability, and accessibility that Robert perceives to be a characteristic of archived information: archived information can be refound. Thus, of his method for creating reminders, Robert says somewhat disparagingly, "I've thought about getting a little notepad you can stick up there [gesturing to the top of his laptop screen], but pretty much my scheme for that is like if I need something short I actually just open Word, and then I usually just save it to my desktop and when I'm done I delete it." By contrast, Robert believes that an archiving tool like Microsoft's OneNote would allow him to label and store these small chunks of information in more permanent, classifiable, and, hence, more useful ways: "Supposedly you can get a lot of stuff in one place . . . you know like I need a little information . . . put into a little document, you know? Consolidate lots of little things I guess was what I was thinking it [OneNote] might do for me."

A compelling explanation for the type of affordance that Robert believes he needs in a memory tool and that leads him to prefer archiving practices over reminding practices can be found in Derrida. In *Archive Fever* (1998) Derrida articulates the concept of consignation as not just a matter of placing solitary items into storage (like saving a document in a file folder) but also as a process of systematizing, classifying, or establishing rich interconnections between all the items in the archive so that nothing is lost and everything is findable (like identifying keywords from an article or attaching descriptive metadata to an image) (Derrida, 1998). Tools like OneNote would allow Robert to "consign" his memories to safekeeping in the archive: "Consignation aims to coordinate a single corpus, in a system or a synchrony in which all the elements articulate the unity of an ideal configuration. In an archive, there should not be any absolute dissociation, any heterogeneity or secret" (Derrida, 1998, p. 3). In a sense then, although he certainly uses *reminders*, Robert prefers not to engage in *reminding* as a memory practice—that is, in creating short-term, single-use, ephemeral texts. Instead, Robert prefers longer-term solutions through which information will not be lost; solutions that are oriented more toward phronesis than techne. As his reference to OneNote reveals, he would like to keep reminders—longer term—in an archive where they can be standardized, classified, and stored in an orderly structure that will render them findable if a future need should arise. A Post-it note on a wall can be lost, forgotten, and rendered unrecoverable once its immediate purpose is fulfilled; a document placed in a database on the other hand leaves traces of its passage. It becomes part of his strategic reserve of information. But intrinsically and practically, creating that reserve of information becomes an archiving memory practice.

A second aspect of archiving that Robert appears to value is that it entails contextualization. Because reminding is, by definition, about creating transitory objects or cues, reminders do not in general need to contain much information about the circumstances surrounding their own creation and instead depend on the circumstances of their use to provide any necessary background context. Further, once reminders serve their purpose as cues, such as a Post-it reminder to go to a meeting at 10:00 a.m., they typically become immediately obsolete. However, the fact that Robert perceived lack of context to be a problem with reminding practices became apparent when he discussed the memory practices of his product team. Robert informed me that, since becoming the semiofficial meeting "note taker" for his product team, he had brought a new rigor to the job of preserving knowledge generated in meetings so that this knowledge would remain useful. Robert contrasted this rigor with the earlier slapdash reminding practices employed by the team during its crucial Sprint planning meetings. Showing me an example of these earlier efforts on the network drive he shares with his teammates, he said:

> Okay, so see what someone has done here is copy the whiteboard. But see there's no [explanation]. That's another thing too: Like a lot of times we can't remember what does 'error this consistent with Mac' mean? [With rising intonation] What does that mean? Whereas with the notes, I just type like crazy, and then I would like write actually more than what would be on the [whiteboard]—like the whiteboard didn't say that. . . . Next time what I'm going to do is cut and paste all the meetings and put them on a separate page that just has meetings, and then what we like to do is get those done early in those thirty days.

Robert prefers the greater contextualization that the activity of consigning these whiteboards to an archive allows him. By transforming the reminders on the whiteboard into written accounts containing a lot of extra details and metadata Robert ensures that what started as, essentially, lists of decontextualized and potentially confusing "action items" become permanent, accessible, and interpretable records in the organizational memory. But this goes well beyond simple reminding practice.

It hardly needs to be said that practices such as Robert's translation of meeting minutes into detailed and organized reports can be quite time consuming and labor intensive. Bowker (2005, p. 116) points out that maintaining useful archives is by nature labor intensive: "The more information you provide in order to make the data useful to the widest community and over the longest time, the more work you have to do. Yet empirical studies have shown time and again that people will not see it as a good use of their time to preserve information about their data beyond what is necessary to guarantee its immediate usefulness."

While the software developers and other members of his software product team—like the majority of us, as Bowker notes—apparently do not believe this extra archiving effort to be a good use of their time, Robert clearly does, and he is not alone in this. This is evident because Robert's emphasis on archiving as a key practice for succeeding at Software Unlimited was echoed by all but one of the other information developers in their initial interviews.[3] Peter, for instance, employs a fairly complex system of legal pads, maintaining one pad per project and rolling important information and action items forward as he fills the pages: "If there's things on that one page that I really want to keep, when I flip it over I'll write them down again." Careful maintenance of this system, Peter believes, helps him diagnose potential usability problems with his product: "I find a lot of what I do is like when people have trouble designing things or implementing things I tend to write that down because that's going to be something a user might have trouble with later if it was hard to design."

The information developer Lucy, who is the focus of chapter 5, engages in a set of archiving practices rivaling Robert's in complexity. The information development manager Becky, on the other hand, offers a sharp contrast to Robert. Becky actually relies on Outlook as her principal archiving tool: "For every project I have a specific email folder so I can file everything. And if it gets too big, I break it down, like 'meeting notes' and different pieces of what I have to do." As do the information developers who report to her, she believes that careful archiving practices both make her more efficient and help maintain her ethos. For example, she asks that each of the information developers email to her daily status reports so that she can keep an accurate and easily retrievable record of their accomplishments: "I find that the writing group as a whole is also very humble, and I think they forget a lot of the great things that they do, so it's nice for me to have a way to kind of go back through the history and check that out." Being able to discover and highlight her team's work contributes to her reputation as a sympathetic and effective manager, qualities noted by each of the information developers on her team.

Each of these information developers believes that his or her archiving practices play an essential role in succeeding at Software Unlimited. They appear to believe that these archiving efforts help them meet the needs of kairos by, for instance, enabling them to contribute valuable insights in meetings, and they acknowledge that careful and consistent archiving helps them internalize the history of the organization, thus contributing to phronesis.

The variety of idiosyncratic strategies and tactics for archiving that each of the information developers acknowledges reveals a generalized memory regime at Software Unlimited that is, as the organization professes, flexible and dynamic, giving even junior employees the tools and freedom to research and organize their information as they see fit. I qualify this conclusion by saying that it represents a "generalized" memory regime because, as I argue throughout this book, organizationally, Software Unlimited features several memory regimes among several roughly corresponding communities of practice. And while Robert's case provides a compelling demonstration of this aspect of a memory regime, it is also important to note that his methods of archiving are, much as we will see with the other members of the information development team, idiosyncratic in nature. Taken together, these idiosyncratic tactics represent all or part of the existing "repertoire of practice(s)"—the variety of discretionary tools—that is the "property" of a given community of practice, which Wenger (1998) talks about, and which we discussed at length in chapter 3.

REMINDING PRACTICES ARE ESSENTIAL TO GETTING WORK DONE

However, as may already be apparent from the foregoing, comparing Robert's account of his memory practices with the action in the observation sessions reveals a significant contradiction—a contradiction that illustrates some less desirable aspects of the memory regime. The contradiction is this: despite the importance Robert assigns to his archiving practices and his corresponding dismissal of his reminding practices, both sessions reveal reminding practices (or, more accurately, the artifacts produced by these practices, the reminders themselves) playing crucial roles in helping Robert succeed in his job duties.

For example, in the first session, while Robert was working on the user guide for his product, he created a Word document in which to write the list of questions he wanted to ask the software developer, Andie, after the Daily Scrum. Before he hurried off to the Daily Scrum, he printed this document and saved it to his desktop, and then, at the conclusion of the meeting with the developers, he offered to leave it with them. These two actions—saving the file to his desktop (the place where he stores temporary reminders— "they . . . find their way to the recycle bin pretty quickly") and his willing-

ness to part with the hard copy after it has fulfilled its purpose—indicate
that creating this document is a reminding practice: the document was
created to serve a single purpose and, therefore, was not intended to live
on in any of the company's official archives after serving that purpose. By
comparison, the portion of the user guide, the "Create a new profile" procedure, which Robert was writing during the same session, would seem to
have been by far the more important document—it was an official text, a
"deliverable" of the type that Robert made sure to store in "a very special
place" on the network drive where "it's all perfect."

Yet the portion of the user guide that Robert worked on during this session was not as visually complex as the reminder document (e.g., it contained no screenshots), nor did it appear to require as much intellectual or
physical effort on Robert's part to produce as the reminder document did.
Furthermore, if Robert accomplished his purpose of convincing the software developers that he was right during their meeting, which he appeared
to succeed in doing, the version of the user guide text he had written that
morning would not have much of a future. That is because, of course, the
procedure would then be substantially revised and improved, replaced altogether with a newer version thanks to Robert's arguments delivered with
the assistance of his reminder.

In other words, despite Robert's initial plans for this work session and
despite conventional expectations that the principal job of a technical communicator entails authoring documentation, the user guide procedure ended
up being the less important of the two documents that Robert produced
that morning. Ultimately, the reminder document—or, more precisely, the
new and essential user-instructional information (information, I might add,
that is in the process of being *transformed* into knowledge)—would, in a
sense, be "transplanted" to and continue to live on in his product's software
interface, while the version of the user guide procedure that Robert created
that morning would probably be superseded and forgotten, the print copy
dropped "in the recycle bin," to use Robert's words, the electronic digital
file "copy" later, presumably, deleted by Robert in due course.

The reminder document succeeded because it responded to the unique
kairos of the situation in a way that a more complete and "bulky" user
guide document could not: it was easily transportable and usable in a hallway meeting, and it spoke to the developers in their own language. Its easy
portability, provisional language, casual appearance, and ample blank space
enabled the reminder document to function as an intermediary between
Robert and the developers, enabling them to translate their ideas in order to
invent a new interface together. In activity theory terms, it was an object of

joint cognition that served to translate the evidence embodied in Robert's own work in his role as a user advocate (e.g., his laborious "nitpicking" of the workflow) into a form that Andie and Gene could appreciate and engage with, thereby helping to win them over to Robert's vision for a more usable product. In short, the reminder document was both suited to the physical circumstances of the meeting and socially appropriate for the audience. Consequently, except insofar as writing the "Create a new profile" procedure occasioned his discovery that there was a problem with the workflow, Robert's documentation efforts throughout this exchange did not produce any text of lasting consequence, while the effort he put into creating an apparently ephemeral reminder document bore fruit (although, of course, that "fruit" would eventually find its way into the "Create a new profile" procedure in the "master" user guide).

Yet, by his own account, Robert does not see the value in this work. He does not recognize the value that is intrinsic to the reminder document itself and does not make any effort to preserve it in his archives (he instead saves it to his Windows desktop, where he places ephemeral documents that he plans to delete). This is further evidence for Robert's conviction that reminding practices are a necessary evil. Despite the vast amount of effort and time it took to create the reminder document, the eventual complexity and sophistication of the document itself, and the insights into the evolving software product that it enabled Robert and the software developers to achieve, both individually and collectively, it remains largely invisible to Robert. Reminding and the texts it produces simply do not loom large in importance in Robert's imagination. He does not recognize the practice as valuable beyond the needs of the immediate task and does not recognize the role that creating it plays in habituating him to the workplace culture and establishing his ethos with his teammates. In sum, Robert's reminding practice—and the knowledge it generates or enables—contributes, in notable contrast with the user guide, to both kairos and hexis/phronesis, yet Robert, in his general disregard for reminding as an important memory practice, recognizes neither contribution.

WHAT ROBERT'S PRACTICES SUGGEST ABOUT THE MEMORY REGIME

The emphasis on archiving and de-emphasizing of reminding in Robert's accounts are indicators of several less-than-ideal and potentially problematic characteristics of the memory regime at Software Unlimited. In fact, on an even more fundamental level, they may point to deeper concerns, and potential dysfunction, within the present organizational structure of the company, which may be the unanticipated and somewhat unwitting result of undergoing rapid growth without adjusting the business model. While Software Limited's organizational growing pains are not the subject of this study, it is necessary to take them into account in order to understand in what ways issues of organizational dysfunction contribute to potential dysfunction in the organizational memory regime, and, on a more molecular level, how these same issues of organizational dysfunction tend to drive— and potentially subvert—the individual memory practices of the organization's members, the employees.

To begin with, these potentially problematic memory regime characteristics affect every employee of the organization, but they can be especially troublesome for newcomers because they influence what counts as important knowledge in the organization, who is empowered to use it, and when exactly they are empowered to do so. First, as would be expected at a software company, knowledge about the company's products and their development history is most highly valued, but this knowledge is not always easy for the information developers to obtain. This is especially true for new information developers like Robert. Second, contributing to the difficulty in obtaining product information, the charge to the information developers to act as user advocates and documentation specialists with their product teams could very easily create an adversarial relationship with the teammates on whom they depend for information. And third, even though they are expected to act as user advocates, the information developers are not provided with the information they need about users to adequately warrant their arguments with the software developers. They lack, in other words, key pieces of knowledge necessary in order for them to attain phronesis.

The influence, as I explain in the following, of each of these three factors significantly affects the information developers' archiving and reminding practices. Combined, these characteristics compel the information developers to meticulously, even obsessively, try to preserve every bit of information—particularly product information—that comes their way. In short, Robert archives because he believes he must (later on, in chap. 5, I will show that Lucy archives because she believes that no one else will, not even the corporate entity itself). Rather than feeling like he is grappling with massive information overload, Robert fears an information dearth, and the only way to overcome this, he seems to believe, is to rigorously consign, classify, and contextualize information. Because they appear not to possess affordances enabling easy systematization reminding practices like his reminder note do not strike Robert as contributing to this ongoing integration as a full member of the organization. They do not, to Robert himself, appear to contribute to phronesis; yet, as I argue above, exactly the opposite is true: Robert's actions of creating the reminders manifest his embodied hexis, the semiconscious habit he has while writing documentation of preparing to warrant and win arguments for better products.

The first potential difficulty facing newcomers and members of communities of practice who do not work directly with the company's products (i.e., who are not software developers in this study) is that much of the mission-critical knowledge that Software Unlimited possesses—much of its organizational memory—does not live in its archives and cannot be found in file cabinets or network drives. Rather, the most important knowledge often lies in the evolving software products themselves, and beyond that, in the individual archives of team members, or in the stories that are told by team members about their evolution, and last, and to a lesser extent, shared within separate communities of practice within the organization. Software Unlimited's products are not, in other words, just the objects of the employee's attention and labor; they are infrastructures that contain, or embody, the company's most important memories. Bowker (2005) makes a similar point when he notes that elements of infrastructures are coconstitutive of the organizations of which they are a part. For example, Bowker notes that, as infrastructures, software programming languages "can have potentially large effects on the nature of the organization. [For example] object orientation is on the one hand a model of the world, and on the other hand the world is learning how to model itself according to object orientation. . . . The programming language is very much part of the organizational history and vice versa" (2005, p. 31). Software Unlimited and its products are similarly modeled on each other. The company's Agile development teams, formed around its products, are simply the most obvious manifestation of this.

In a sense then, the products are themselves the organization's memory, and invoking knowledge of these products, including the subtlest nuances of their use and purpose as well as the circumstances of their history, is the surest way to win arguments focused on their future innovative evolution and development. The second observation session of his memory work, Robert's meeting with Wallace, provides a good illustration of the power of product knowledge to trump other warrants. Where Robert appeals primarily to the experience of the user to warrant his arguments, Wallace appeals almost exclusively to his superior knowledge of the product and its development history. For example, when Robert attempts to argue that the issue will confront users with a novel situation, Wallace counters, "This is not really a new thing." In other words, based on his longer history working on the product, Wallace knows that it is not a new problem. He is asking, essentially, "If users haven't complained before, why will they complain now?" Robert must concede this and other points because he does not and cannot know this history, and, in the end, Robert is only able to enlist Wallace in cause as a tepid ally at best: In a follow-up email a few months after this meeting, Robert noted that Wallace "probably sees my point of view 75%, but is not nearly as passionate and has other problems of higher priority." Ultimately, then, in both his initial resistance to and eventual (grudging) acceptance of Robert's critique of the product interface and workflow, Wallace primarily relies upon his superior knowledge of the product and its history to defend the status quo.

Thus, the chief difficulty for newcomers like Robert and employees with less direct access to the software products themselves is that, unlike knowledge held on a network drive, which can be made accessible to all employees relatively easily, knowledge held in product infrastructures is continually evolving and growing as the products themselves evolve. The problem is that not all employees possess equal access to or understanding of these evolving products, and those who do possess knowledge appear complacent about consigning the knowledge to archives accessible by non-specialists (i.e., they do not create written accounts of design decisions or of the reasoning behind these decision). Robert's account of his efforts to classify, contextualize, and properly archive the whiteboard detail mentioned earlier demonstrates this: the other members of his product team were content to carry out the design changes decided on in the meeting without worrying about preserving any detail of the reasoning behind these decisions. In other words, they assumed that the software itself could explain the necessary context. As Bowker puts it, "You don't need a memory if evolution is doing your thinking for you" (Bowker, 2005, p. 100).

As a consequence of this, new employees like Robert are forced to spend a great deal of time and effort locating, classifying, and contextualizing information about the company's products in order to render this information useful to their work. When Robert notes that "a lot of times we can't remember what does 'error this consistent with Mac' mean? What does that mean?" the "we" to which he is referring is not the software developers; it is the information developers. They are the ones who need this information. Robert and the other information developers must, in short, perform a great deal more work than the software developers if they want to win arguments. No wonder, then, that when they are asked to talk about their memory practices, the information developers think in terms of archiving—of fixing, stabilizing, or otherwise preserving memories for the long term. That is what the memory regime requires them to do. Robert summarizes this best when he points out that "it's a very [software] developer-centric work environment." What counts as memory for software developers counts for everyone—or is implicitly expected to count for everyone.

The second difficulty facing the information developers, particularly the newcomer Robert, arises from the fact that their user advocacy role can create tension with those members of the organization on whom they rely most for information, the software developers, whose memory, as just noted, is expected in some very significant ways to count for everyone. As I noted earlier in this section, at Software Unlimited, the duty of advocating for users during the software design process is part of the information developer's job description. This places the information developers in a potentially adversarial relationship with the other members of their product teams from the start because, as Richard Buchanan (1989) argues, the design of any new technology is inevitably a politically charged activity entailing dialogue, dispute, and consensus seeking about preferable courses of action in situations without obvious solutions. Any change to the evolving software product means extra work for the team—especially the software developers—and may also lead to long nights, added stress, hurt feelings, and bruised egos.

Here it is helpful to take a step back and explore how, as I also suggested at the beginning of this section, a metastasizing organizational flaw may be partly at fault in the creation of this kind of adversarial atmosphere. Later, in chapter 6, the six-year veteran information developer, Angela, explains that the nascent origin of the user advocacy function, as subsumed under the Information Development Department was—Angela herself. We will go into greater detail in chapter 6, but in brief, Angela was originally hired not as a technical writer but as a process improvement manager—essentially, to make the company's software products user friendly—and as part of that function, she became, if informally, the company's first user advocate. The test of successful user compatibility came to be whether one could answer yes to the basic, simple question, "Can Angela use it?" Later, Angela became the company's first information developer, and one may easily connect the dots.

Now, that somewhat incestuous (for lack of a better term!) business operational model may have worked very well (as Angela insists it did) when, on the one hand, Software Unlimited's product offerings were limited enough for one individual to be, for all practical purposed, the sole beta tester for all products. And on the other, it may also have worked well when the company consisted of fewer then twenty employees to whom a sole beta tester could easily report the results, good and bad, of her product test drives, all in a single convenient forum by gathering everyone together in one room.

But the company at the time of this study is now 120 people with numerous product lines, all of which are undergoing constant, even furious upgrades, the addition of new features and applications, and so on. Further compounding this situation, in response to this expansion the company has also created a new UX department, whose primary responsibility is user experience design, thus further obfuscating the user advocacy identity of the information developers. I use the term "identity" instead of "role" here deliberately: If the software developers do not see and "identify" the information developers as having an authoritative role as mediators of user advocacy (and if they further see and identify a whole separate, dedicated UX department as being charged with this role), they will be even less inclined to pay any attention to user interface concerns raised by the information developers. In fact, in chapter 6 I report a series of disagreements, ultimately resulting in a confrontation of sorts between Angela and Carl, a member of the training team, over just this issue of who is responsible for championing user advocacy. The conclusion that one may draw from all of this seems clear: The rapid expansion of the company, and its current size, may well mean that this organizational model, in which the technical information writers are expected to act as advocates for technical information users, no longer works at Software Unlimited.

Concomitantly, the fast-paced Agile software development environment, recently implemented at Software Unlimited, only exacerbates the potential for conflict. On one hand, the organization values the careful research necessary to make informed user-centered decisions, but on the other, speed and forward momentum are the real motivators: any information that threatens to slow the development process—as well as the bearer of that information—may be perceived as irrelevant or counterproductive. Robert must negotiate carefully if he wants to succeed in his efforts to steer the development of his product. Thus, even though user advocacy represents an ethically desirable and potentially status-enhancing role for technical communicators to play on design teams (e.g., Salvo, 2001; R. R. Johnson, 1998; Johnson-Eilola, 1996), the role of user advocate may make their memory work more difficult.

This tension is tellingly illustrated in accounts of Robert's research and information management skills. While the information developers appeared to have already grown to value Robert's research skills and information managing abilities quite highly, Robert indicated that the members of his software development team had not. In fact, at several points in our initial interview Robert indicated that he had experienced difficulty in learning from and integrating with the other members of this team. For example, Robert offered a far more reticent and cautious image of himself as a researcher in our interview than his information developer colleagues presented in their accounts. In particular, Robert noted that he had had to learn several lessons "the hard way" since coming to Software Unlimited and expressed a wish for a more orderly and formal method for learning about the company's products and development processes than that offered by the company's (apparently) cursory newcomer orientation class.

Robert indicates that at least one of these hard-learned lessons resulted from overstepping his newcomer status by assuming he knew more about how the organization operated, an attitude that may have been perceived as presumptuous by his more senior colleagues on the product team: "For being new, I don't know who's good at what, who's assigned to what. So, at first I never really asked questions, I just kind of assumed." Robert learned his lesson and subsequently paid more attention to building closer relationships with the other members of this particular community of practice: "Now I go find out who's the best—who's going to solve problems. So I try to make friends with all the programmers who are doing the website or [other product design work]. Now it's actually getting better."

As his meeting with the software developers Andie and Gene in the first observation session demonstrates, Robert has learned that discovering the information he needs and getting the results he wants at Software Unlimited require carefully maintaining relationships with his teammates, particularly the software developers. I would again suggest that this contributes to Robert's preoccupation with obtaining the backstory, the context, and the extra details that he believes are critical to his success—all, in other words, contribute to phronetic knowledge

The third aspect of the memory regime that complicates the memory work of the information developers is that knowledge about users of the type that the information developers must invoke to warrant their arguments can be difficult to acquire and use. Like product knowledge, knowledge about the organization's users is inaccessible to the information developers. Despite the information developers' charge to act as user advocates in design processes, and despite Software Unlimited's dedication to the user experience (usability testing and product demonstrations for customers were a regular occurrence during the period of the study), there was no central repository for user testing data, nor were the information developers themselves empowered to conduct usability testing on their own. When asked if there was some central source where all this usability-testing data lived, to be available to the information developers to make their arguments, the information development manager Becky answered in the negative: "I think you just have lots of little sources for it when you are doing your work. It would be amazing to have that sort of one-stop place to get all that information for research or whatever, but I just don't know that anyone has that as their job."

An example of one of these dispersed "little sources" can be found in the account of the second observation session. Recall that at the beginning of the scene Robert informed me that he had become alarmed at the complexity of his products' interface because of comments posted by a user to the product's beta-test wiki, which Robert himself moderated. These user comments served, therefore, as the initial impetus for what turned out to be a lengthy and time-consuming campaign to change the integration workflow. But, despite all of his otherwise elaborate preparations and careful warrants, Robert never directly references this user's feedback when presenting his case, either in his PowerPoint presentation or in his oral argument with Wallace. Instead, Robert creates and then impersonates a fictional user persona who will "get so mad" if the interface remains unchanged.

The relatively dispersed state of the company's knowledge about its users and the unavailability of direct access to direct user testing information were indicated in other ways as well. For instance, when asked if user testing data were made available to them, the information developer Monica noted that, while she and the other information developers were free to watch the user testing as it occurred, more permanent records or detailed studies were not made generally available but, instead, had to be acquired from the team conducting the test: "They show us an overview, and if we wanted to see more details or if I had specific questions about something, I could go and ask someone to show me their findings and research or data." In fact, the issue of which teams were permitted to conduct user testing became a source of controversy during the period of the study. When the information developers conducted their own user test of a particular product, that product's design team contested both the information development team's findings and its right to conduct the tests. Apparently cowed by this challenge, the information developers did not attempt to conduct any further user testing on this or any other product during the study.

Thus, lacking full access to the histories of either the products or users' interactions with those products, information developers, particularly newer ones like Robert, must perform a great deal of memory work to achieve phronesis. Often this work takes the form of efforts to in some fashion stabilize, contextualize, embody, and otherwise improve the durability of knowledge about users.

These are troubling findings for Software Unlimited, because they reveal that crucial user interface information obtained through a variety of sources, from internal and outside beta testing to actual user reports (perhaps the most critical information the company possesses) is scattered widely across the various the various departments and individuals working on different product lines with no distinct and easily accessible repository. As a direct consequence, product design and development, product improvement and innovation, and product integration and optimization are all—as we have seen—negatively affected. This is because no one individual or set of individuals, no one particular department, and indeed, no one specific community of practice as a whole is charged with acquiring and possessing all of the information, beta test results, real-time user reports, or other data relating to the two most vital (and sensitive) categories of information that are most critical to the company's success: (1) user interface concerning consumer customer ease of use and satisfaction; and (2) product enhancements and efficiency upgrades designed to correct newly discovered flaws, or more generally to adhere to high quality standards for product functionality and transparency of use.

What Software Unlimited possesses instead are bits and pieces of discrete, largely random information stored in and across a plethora of different, haphazard, unconnected data sources, or otherwise "owned" (and not directly or automatically shared) by different individuals on different product teams with no overarching rationale. Further, as we have seen, and which also is probably inevitable given these circumstances, territoriality rears its ugly head in the form of the beginnings of turf wars over jurisdiction—specifically, who or what departments are responsibility for acquiring specific critical information on the one hand, and who is entitled to "ownership and use" of that information on the other (I explore a prime example of this in the chapter 6 case study of Angela). As a case in point, in Robert's observation sessions, we see a virtual tug-of-war both over the crucial informational concerns that Robert brings to his meetings with the software developers and over who should prevail in dictating which specific, significant "fixes" to product configurations should be implemented in response to those concerns.

The potentially ominous implications of this set of circumstances should not be ignored. In the highly competitive computer software industry, the terms "user interface" and "user advocacy" are, practically speaking, essentially euphemisms for customer service. If Software Unlimited were to continue in this manner to play fast and loose (literally and figuratively) with issues of user interface and user advocacy through the product development and rollout stages, so that some of these potential user interface flaws actually made their way into the units sold to customers (a situation I will look at in chap. 6), the company's reputation in the marketplace would suffer. As I stated earlier, it is not the focus of this study to evaluate the corporate workings of Software Unlimited; still, any upper-management individuals informed of this situation have reason for concern.

More specific to my focus in this book, this state of affairs further supports my contention, stated in chapter 3, that Software Unlimited does not possess or operate with a strict, centralized, and hierarchical memory regime but rather possesses and operates with a number of smaller and overlapping "satellite" memory regimes within its overall corporate structure. Furthermore, the situation also serves to further our understanding of why most of the information developers (and, one may presume, many of the software developers and other employees) are so anxious to collect and archive as much information and knowledge as they can get their hands on. With vital information such as user interface data and product integration data so widely scattered about the Software Unlimited universe—and not collected in a single-source, centralized, and accessible database anywhere in that universe—every single shred of information that Robert and the other information developers can capture is intensely valuable, at least potentially. In a sense, then, the conditions currently prevailing at Software Unlimited make phronesis ultimately unobtainable, at least for the information developers: They may be excellent writers of software documentation (i.e., masters of the techne of technical writing), and they may be regarded as full members of their communities of practice, but, as things stand and despite their most strenuous efforts, they remain unable to achieve the phronesis necessary to act with full knowledge in the social situations presented in their workplace.

CONCLUSION

These scenes of Robert's memory work have begun to delineate some of the details of the memory regime at Software Unlimited. Understandably, given that it is a software development company and earns its profits by selling its software products, this memory regime appears to regard knowledge of its software products and the circumstances and decisions leading to their current iterations as most important. However, this knowledge, although, in a sense, universally "visible" to everyone in the company via the product interfaces, is nevertheless unavailable to many employees, because the evolving product interfaces are alone incapable of conveying information about the circumstances and contexts of their own creation in accessible language. The information developers, particularly newer ones like Robert, then, must expend a great deal of effort on memory work attempting to find, stabilize, and contextualize this knowledge, and then to place it into a system where it can be readily found if needed. Unfortunately for the information developers, this knowledge is often available only from the software developers, who may not be willing to part with information that can create more work for themselves.

In addition to product knowledge, these sessions with Robert also revealed user knowledge to be important to this organization, especially to the information developers, for whom user advocacy is a principal job responsibility. However, because, as with the products, the information developers do not possess direct, unmediated access to this knowledge, they must spend a great deal of extra time and effort attempting to uncover and then fix this knowledge in some way in order to use it to warrant arguments and win allies. It was precisely in this effort of trying to translate this domain of knowledge so that it could influence the behavior of his coworkers that Robert spent so much of his attention and effort. The limited measure of success he experienced in these efforts is some indication of the obduracy of this memory regime.

In seeking answers to the research questions that motivate the study, this chapter has begun to demonstrate for researchers the value of reconceptualizing information-managing work as rhetorical memory practices: Robert, like most contemporary knowledge workers, stays on top of information via a wide array of practices, but such practices change both the information and person managing that information. From observing Robert and his practices, we have been given a first indication of what a contemporary art of memory might looks like. The next chapter, which focuses on Lucy, will begin to show how collective and emergent memory practices of memory regimes are manifested in embodied behavior.

CHAPTER FIVE

Embodying Memory

> That memory and recollection could best be understood as a physical
> process involving physical organs is both fundamental to the whole idea
> of memory *training,* and quite foreign to much post-Cartesian episte-
> mology.
> —Mary Carruthers, 1998, p. 48

This chapter focuses on Lucy, an experienced technical writer struggling to balance caring for her growing family with new duties and responsibilities to the firm. Lucy has been employed at Software Unlimited for two years and, therefore, possesses a fair degree of seniority in the fast-growing company. However, Lucy's tenure with the firm has not been a conventional one: two months after she was hired she went on maternity leave, but, less than a month after she returned from leave, her older child was diagnosed with autism and she shifted to part-time work so that she could care for him while he began therapy. Only in the month prior to our interview had Lucy returned to working full time and to serving as "documentation lead" for the development team working on Software Unlimited's most complex product. Lucy's case, therefore, offers us an opportunity to explore how overextended knowledge workers employ multiple memory practices to help themselves achieve work-family balance.

Like Robert, Lucy belongs to two communities of practice within the organization: the information development team and a software product development team. Also like Robert, Lucy finds the information development team supportive and appreciative of her knowledge and skills. In part, this support and appreciation stem from Lucy's educational and employment background—she earned a bachelor's degree in technical and professional communication and worked as a technical communicator for a large company for five years before coming to Software Unlimited. Additionally, Lucy's two-year affiliation with the company makes her second only to Angela in seniority among the information developers, even though she has worked little of this time as a full-time employee on a regular schedule. As a consequence of these factors, the other members of the information development team appear to regard Lucy as a full, even as a senior, member of the team. Evidence of this regard can be seen in the fact that, in spite of her part-time status, she was asked to mentor newer employees, like Peter and Robert, when they joined the company. For his part, Peter believes he has learned a lot from her: "she's really great at being organized. . . . When I worked with her it was enlightening to see how she worked."

Overall, Lucy appreciates the freedom offered by the memory regime at Software Unlimited in comparison to her previous employer, but, like Robert, she indicates that the lack of structure made things challenging for her when she was new: "It was very disconcerting for me when I started working here that I didn't have templates, that nobody had templates they were giving me. I considered that I could work pretty independently—and I could—but it was really much more than I thought it would be." That is, even though Lucy did not always appreciate the strictures of the memory regime at her earlier job—"I remember complaining about them when I had them"—the rigorous reporting requirements, mandatory nightly backups of local information, abundant templates to assist composing, and close managerial oversight of her work tasks had conditioned her to not having to think about information management issues.

Despite her initial discomfort at the more laissez-faire aspects of the memory regime of Software Unlimited, Lucy has gained confidence in her systems for managing information and staying organized. She has, she emphasizes, become "really proactive" and self-reliant in her memory practices because "no one else is going to track it." The respect of her information development team colleagues would appear to attest to the success and thoroughness of her practices.

Lucy's status and ethos are not as well established with her software development team, however. First, in spite of her two years with the company, few of these teammates had worked with her before she was appointed documentation lead for their product (Becky had served as the information lead for the team for more than a year before being promoted to team manager). In other words, although Lucy may not be a newcomer to the organization, she *is* a newcomer to this particular community of practice and must earn her place accordingly. Further complicating her relationship with the other members of her software team, since returning to full time work, management has permitted Lucy to work a staggered daily schedule from 6:00 a.m. until 3:00 p.m. so that she can pick up her children from daycare and school. The combination of her unusual history with the firm, her current staggered work schedule, and the general concerns entailed in caring for a busy young family conspire to make Lucy keenly anxious about her teammates' perceptions of her contribution: "Since I came back, I feel a lot of pressure. . . . I feel like I'm the most restricted person on the team in how much I can give."

Although her situation is atypical at Software Unlimited, Lucy's is becoming an increasingly common scenario in the early twenty-first century, particularly among female workers with small children, 58 percent of whom now work outside the home (U.S. Department of Labor, Bureau of Labor Statistics, 2011). Further, even though Lucy is a highly skilled white-collar professional working for a progressive employer in an industry with a good track record of offering family-friendly benefits, achieving work-family balance remains difficult. Research indicates that, even though they offer flexible work hours and other family-friendly policies for highly educated employees like Lucy, high-tech firm like Software Unlimited often put considerable pressure on employees to work long hours and to value production deadlines over family obligations (American Association of University Women, 2003). In such "high-commitment" firms, employees can feel considerable pressure to put in "'voluntary' long hours and [a] willingness to subordinate other aspects of one's life to the demands of work" (Poster & Prasad, 2005, p. 126). Lucy appears to be expressing a similar worry when she says, "The thing is, it [family caregiving] affects your work life, but at the same time they can't take it easier on you or anything because you have kids. So, it gets really hard. Because I want people to know that I can do the stuff. Even if I'm not here I can work at home, but I get the feeling there's this impression that if you aren't here [you aren't putting in as much effort]."

As a consequence of the pressure she feels, Lucy believes that she must attempt to segment and regiment every aspect of her working life "in ever more inventive orders" (Geisler, 2001, p. 321). To do this, Lucy takes advantage of most of the scheduling and information-managing tools that the firm offers and is continually searching for new tools and solutions to help her stay even better organized and focused. These efforts have made Lucy, in both her own and the other information developers' estimations, the most carefully organized and regimented member of the information development team: Angela is principally thinking of Lucy when she notes, "There are several people on the team who are very, very structured. They like lots of lists and check lists."

With regard to Lucy, then, it appears we are dealing with an individual with midrange experience and expertise who, in high contrast to Angela, is much more comfortable in utilizing the memory practices of finding, archiving, reminding, and referencing—perhaps at times to an excessive degree, as we will see. Furthermore, in doing so, Lucy relies to a large extent on the "physical" affordances of the office environment such as computer interfaces, information-sharing databases, and document storage for later retrieval—and to a lesser extent on office machines like photocopiers, printers, filing folders, and cabinets for storage of paper documentation.

Analyzing memory work centered on Lucy thus offers, I would suggest, important insight into how today's overextended and information-saturated knowledge workers leverage every memory-assisting affordance offered by their embodied contexts in order to track and record every shred of incoming (and outgoing) information based on the deep, sometimes vexing concern that it could turn out to be critical to their information development role, to one or more of their communities of practice, or to the organization as a whole and, consequently, crucial to the quality of their job performance and their achievement of phronesis. I emphasize embodiment because, as situated-cognition theory points out, knowledge workers placed in high-pressure situations like Lucy's often feel that they must learn to use every tool and technique that they perceive to be available in their embodied (physical and digital) contexts to help themselves do the memory work of managing information: "Learners . . . always lean on whatever context is available for help" (Brown, Collins, & Duguid, 1989, p. 36). In other words, it is often the case that successful knowledge workers offload or delegate much of the burden of memory to their surrounding tools and environment. This practice of offloading, in practical terms, would appear to be a positive one given the conditions of extreme information overload that knowledge workers, and technical communicators in particular, face in today's world. However, as Lucy's case may illustrate, there is a distinct danger for duplication of both information and the effort to manage that information. Specifically, powerful project planning and scheduling programs can overlap; similarly, the ease of creating new databases or of populating such databases with multiple "updated" versions of documents and manuals can make it difficult to keep chronological track of "most recent" versions in a centralized way. New or novice information managers, faced with the constant and unrelenting influx of pushed information, both from outside and within the organization, and worried about cataloging all of that information for fear of missing some bit that might prove critically important are apt to compile large amounts of data in a manner akin to overstuffing the file cabinet in the days before computers.

In offloading memory in this way, not only do knowledge workers shape their tools and environments, but their tools and environments tend to shape them as well. Recall if you will the earlier metaphor of the airline pilot who must himself, in essence, become "programmed" to pushing buttons, flipping switches, pulling levers, and handling the controls in precisely the correct order and degree that are critical to flying the craft safely and successfully, or the athlete who must program every physical aspect of her body into a strict sequence of physical movements and eye coordination to achieve a high percentage of shots made. In this way, therefore, Lucy's case offers us a chance to see the memory regime reflected in Lucy's embodied disposition or hexis.

FIRST OBSERVATION SESSION: BUILDING THE NEW USER GUIDE

When I arrive at Lucy's cubicle at around 9:00 a.m., Lucy informs me that her goal for this work session is to continue working on a new user guide that she is creating for the next release of her product. The purpose of this particular user guide is to help people who are new to the product, so Lucy wants to make the guide as short and streamlined as possible. As source material, Lucy is using the large Word document that serves as the source file for the online help for the current version of her product. Her process is to copy and paste text from this online help document into the new user guide, shortening and simplifying the pasted text to make it more appropriate for beginners, updating the terminology and procedures where necessary to match the new release of the product, and adding helpful notes giving readers suggestions for online help topics that they may consult if they need additional information.

Lucy begins work by browsing in Windows Explorer to find her working document. This document is an in-progress draft of a chapter of the beginner's user guide that will be included with the new version of Lucy's product. Lucy quickly browses to the Sprint folder she is seeking in the chronological list of Sprints. She opens this folder but then pauses to read the names of the individual files in the folder. As she reads, she moves her cursor down the list in a manner similar to running one's finger down a list on a piece of paper.

Lucy appears unsure which chapter she wants to open, because, rather than opening one of the files in the list, she opens her Outlook calendar, finds the calendar entry for the current block of time, which is labeled "Work on Method.doc," and views the details for the time block. This cal-

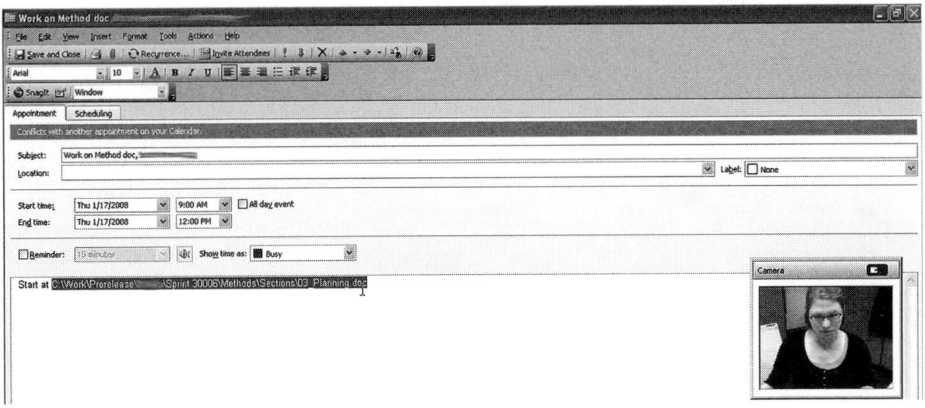

Figure 5.1   Lucy copying a file path from a calendar entry.

endar entry contains the text "Start at C:\\Work\Prerelease\[product]\Sprint 30006\Methods\Sections\03_Planning.doc" (see fig. 5.1). Lucy copies and pastes this path into the address line of the web toolbar in Word in order to open the file.

After it opens, Lucy scrolls through the document until she arrives at a block of text highlighted in yellow. (In our initial interview, Lucy had explained to me that she uses Word's color highlighting such as this one to remind herself where she needs to start working, adding that she prefers the highlighting feature over Word's Comment feature because highlighting enables her to select multiple custom colors for different reminding purposes.) She removes the yellow highlight from the text and begins to read the text for the next several seconds. As she reads, Lucy moves the cursor over each portion of the text as she comes to it.

After working on this document for about two minutes, Lucy apparently realizes that she needs another document because, leaving a gap in her text in progress, she returns to Windows Explorer and begins again to browse the folder structure on her hard drive. She navigates up and down the Sprint file structure until she finds the file she is looking for, a process that takes about thirty seconds. As she browses, she leans in toward her monitor, bringing her left hand to her face and partially resting her head on her hand, a position she holds until she finds and opens the document she seeks (see fig. 5.2). She repeats this gesture four more times during the session while browsing or reading. Each time she does so, she is either browsing for information or reading extended pieces of text. The movement apparently indicates that she is dedicating extra attention to onscreen information.

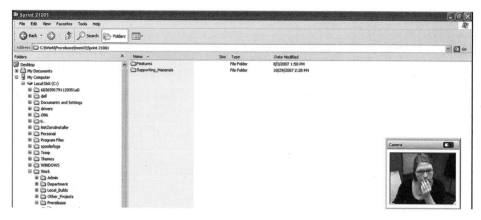

Figure 5.2 Lucy pensively browsing through the files on her hard drive.

After finding the online help document from which she plans to copy content for the user guide, Lucy opens it. Next, in order to find the text she needs, Lucy opens the Find and Replace dialogue box and searches for a specific word. This search takes Lucy to the section of the large document that contains the text she needs. After briefly scrolling through the section, she locates the text she needs and copies and pastes this text into her first document. After copying the text into the document, she revises the wording of the pasted text to reflect the new context and new version of the product. She repeats this find-copy-paste-update process several times over the next fifteen minutes.

At this point, Lucy appears to realize that she needs a third document, because she abruptly ceases typing in the middle of a sentence and begins trying to find a file by browsing in Windows Explorer. This file appears to be difficult to find, because Lucy spends more than forty-five seconds browsing the folder structure looking for it (during which time she again rests her chin on her left hand) and because she opens and then promptly closes what appears to be an incorrect document before finally finding the one she is seeking. This newest document is only three pages long, and Lucy locates the section she wants by scrolling. She intently reads the section for several seconds, again moving the cursor over each section of text as she appears to read it, before returning to the first document containing her text in progress. She proceeds to take out the gerunds from the section she is writing. (She later informed me that she opened the third document to use as a style guide.) Although she leaves it open for the remainder of the session, this is the only time Lucy uses this third document.

Lucy continues working as before, switching between her working document and the help compilation document, until, at 9:55 a.m., an Outlook calendar pop-up reminder appears, reminding her that her Daily Scrum will begin in five minutes. Lucy performs several steps to preserve her job state and prepare herself for the next time she works on these documents. She first highlights the section she is working on in the user guide, apparently so that she will know where to begin work next time. Next, she inserts an MS Word comment on another section with the text "does this need an intro section?" (she later informed me that this question was for herself). Finally, she saves the file and departs for her meeting.

SECOND OBSERVATION SESSION: SEARCHING FOR A SOLUTION TO A SOFTWARE PROBLEM

Approximately one month after the first session, Lucy arranges for me to again observe and collect data. Her goal for this session, she informs me, is to finish creating a Word document that will become the source file for the online help of a new feature in the upcoming release of her product.

Lucy begins by attempting to find the Word document that she plans to work on—the document that serves as the source file for the online help of a new "Export to Word" feature of her product. As she did during the first observation, she starts this finding process by browsing through the contents and subfolders of the "Prerelease" folder on her local drive using Windows Explorer. She does this for about a minute without success before using the search function of Windows Explorer. This search also fails to return any results, so she resumes browsing. After about four minutes spent in unsuccessful searching and browsing, she selects a help file source document for a similar feature, "Export to PowerPoint," opens it, replaces all instances of the word "PowerPoint" with "Word" using the Find and Replace function, and then selects Save As, renaming the document for the new feature.

She then opens Quadralay WebWorks, the help-file-generating software used by the information development team at Software Unlimited. She creates a new project in WebWorks, selects the "Export to Word" document she has just created to be the source file for the online help project, and prepares to save the project file. In the Save dialogue box, she first browses to the folder of the most recent Sprint, "Sprint 3006." After pausing for a second, however, she navigates one level up to the folder "Prerelease" and creates a subfolder for a new Sprint, "Sprint 3007," into which she saves the project. While she waits for her help file to generate, Lucy spends several minutes updating her Outlook calendar. Interestingly—and in keeping with the role that the Outlook calendar plays in helping her preserve a record of her activities—she retrospectively updates one event that has already occurred by expanding the planned duration in order to, presumably, more accurately reflect the actual time the activity took to complete.

When it finishes generating, Lucy opens the online help file. She appears to almost instantly recognize that something is wrong with the newly generated file, because she immediately begins browsing through the menus of the WebWorks application. She is apparently unable to locate the information she is looking for in the menus, so she opens the WebWorks online help. While waiting for the online help to open, Lucy selects another option in the application, which opens the WebWorks wiki in a new browser tab. She scans the list of hyperlinked subtopics on this wiki page for about twenty seconds, before switching to the browser tab in which the online help has opened. On the online help tab, Lucy performs a logical text search for a particular term. This search returns about twenty topics, ranked in descending order of match. She quickly opens the first topic in the list and proceeds to read it for about one minute. The help topic is text heavy and contains screenshots, and Lucy reads it with some care, moving the cursor over the text at various points as she reads and scrolling down and then back up several times. Apparently dissatisfied with the information the topic provides, she opens two more topics in quick succession, scanning both quickly. Altogether to this point, Lucy has spent almost five minutes trying to solve the problem. She has also opened nine topic or help files or "pages" in the process (not including her aside to make updates to the Outlook calendar).

Lucy next opens the WebWorks website, where, after scanning the home page, she moves her cursor over the entries in the left navigation bar to deploy pop-up submenu options, eventually selecting the FAQ page from one of these pop-up menus (see fig. 5.3). She scans this FAQ page for about one minute before switching back to the WebWorks application, where she continues to explore her project settings. Next, she opens the WebWorks wiki where WebWorks users are able to contribute to the general store of knowledge about that application, but again she appears unable to locate the information she needs, though she both browses and searches the site. She continues switching among these four sources—the two websites, the online help, and the application settings—for the next ten minutes.

At this point, Lucy leaves her cubicle and goes to consult fellow information developer Monica in her office to see if she can help solve the problem. Monica informs Lucy that she is having a similar problem finding some

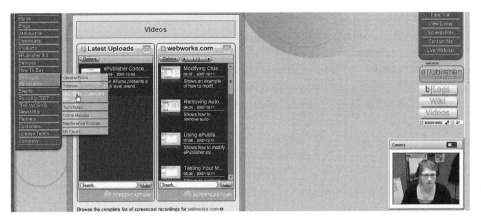

Figure 5.3 Lucy browsing the Technical Assistance submenu of the WebWorks website.

critical information herself and suggests that they meet on a future day to discuss Lucy's problem. Lucy returns to her cubicle, where she spends the next forty minutes alternately searching the websites, adjusting her project settings, tinkering with the Word document, and regenerating the help file. Lucy spends an extended portion of this period reading and scrolling up and down one particular wiki page, which appears very technical and contains several step-by-step procedures. Lucy reads this page very carefully before performing one of the procedures in an attempt to fix her document. While following this procedure, she carefully references the wiki page by switching between the browser and Word at each step.

Still apparently unable to resolve her problem, Lucy eventually bookmarks the wiki page, placing it in her browser bookmark folder "Doc Tools." She then names or labels it so that the name describes the problem she is attempting to solve. She saves the Word document to the N drive, appending "_draft," to the file name, and sends an email and hypertext link to this file to a software developer, asking him to review its contents for accuracy. Finally, she concludes this activity by shutting down WebWorks, taking one last look at the wiki page, and closing her browser.

LUCY'S MEMORY WORK

In these two sessions, we observe Lucy engaged in all of the following activities, each of which may constitute forms of the memory practices that I identified in chapter 2:

Browsing the folder structure on her computer and the network hard
    drive, attempting to locate various documents, a *finding* practice

Performing text-based searches within large documents in order to locate
    specific sections within those document, another *finding* practice

Employing a color-coding scheme within her working document to re-
    mind herself of the status of the text the next time she opens it, a
    reminding practice

Viewing the Outlook calendar entry for a work task in order to find the ·
    location of the document she plans to work on, another *finding* prac-
    tice

Creating new document files to store, ostensibly more conveniently, in-
    formation that she appears to have tried to find for an extended time
    and had difficulty finding, an *archiving* practice

Consulting with another information developer in order to seek help
    resolving a technical issue with the online-help-generating software,
    a *referencing* practice.

What is most striking about Lucy's working routine is her heavy reliance
on the computerized information and knowledge-based affordances available to her within the confines of her own office cubicle. Lucy's method of
operation thus stands in stark contrast to Angela's; whereas Angela relies
almost exclusively on what is "in her head," Lucy appears to do almost the
exact opposite through her reliance on the internal data systems of Software Unlimited. In only one brief instance does Lucy seek help outside the
confines of those systems or even, physically, beyond her cubicle, when she
attempts to consult with a fellow information developer, which is perhaps
the closest Lucy comes to Angela's "holler around the corner" way of informally consulting with coworkers on technical or operational issues. It
may be instructive to add that Lucy's brief consultation with Monica does
not result in a resolution of the issue, with which Monica has also been
struggling, but rather an agreement to meet at some (undetermined) time
to discuss the problem. The disparity between Lucy's and Angela's work
methodologies demonstrates that Lucy relies on an entirely different set
of memory practices from those of Angela, who, as we will later see, relies
almost exclusively on the memory practice of storytelling.

Moreover, Lucy's embodied approach to her memory work offers an
opportunity to consider the ways in which situated-cognition theory can
inform the study of information management. At a minimum, it offers
some insight into the process of the transformation of raw information into
knowledge for product-based applications by and among the information

developers at Software Unlimited. In this context, it is important to keep in mind that among Lucy's primary objectives, by her own thinking, are to learn the company systems—in effect, its fundamental memory regime (or its multiple memory regimes)—and to effectively and visibly demonstrate to other members of her teams that she works as diligently and as productively as everyone else despite her unique and nonconventional employment schedule. Thus, both individual learning and community participation are keys to success for Lucy.

With respect to the learning component, Reynolds, Sinatra, and Jetton (1996) state that situated cognition "attempts to account for how one learns in a conceptual environment. The conceptual environment consists of the external world as perceived, the internal representations of the perceptions, and the resulting interactions" (p. 100). Reynolds, Sinatra, and Jetton go on to state that situated cognition emphasizes not simply "the processes that occur in the mind," but, more importantly, "the affordances within the environment that contribute to the formation of mental models" (p. 101). And finally, they contend that "the agent . . . perceive[s] affordances in the environment . . . through the frequency of engagement in activities" (p. 101).

Just as important for Lucy is that she feel that she is a valuable, effective, and competent member of her team and the organization. Barab and Plucker (2002) cite Lave and Wenger's (1991) eloquent formulation of why learning is not just important for its own sake, but is crucial as an aspect of all activity within an organization to ensure "legitimate peripheral participation" of individual members in communities of practice. Barab and Plucker (2002) state:

> In their view [Lave & Wenger, 1991], learning is not simply one kind of activity, rather learning—or, alternatively, *talent development*—is viewed as an aspect of all activity.
>
> The process of learning, which is always situated, must be described in relation to the context through which it occurs. Of prime importance are one's reasons for learning. This line of thinking led to the notion of legitimate peripheral participation, in which the primary motivation for learning involves *participating in authentic activities that move one towards becoming more central to a community of practice.*" (173; emphasis mine)

Returning to the memory practices observed in the two sessions with Lucy, we may first note that Lucy perceives all of these activities to be critical to learning and developing proficiency with the company's memory

regime, to managing the continual updating of product knowledge, and to
her efforts to be effective and timely in her role as technical information
writer and communicator. Like Robert, Lucy engages in these practices
within the context of the evolving memory regime of Software Unlimited:
existing methods of preserving and using information influence Lucy's practices while her practices, if successful, simultaneously influence and change
the regime. Now that we have observed Lucy's memory work, we will want
to explore and understand the kinds of memory practices that she most relies on, but we can also begin considering her case in relation to the study's
research questions. In particular, Lucy's case offers us the opportunity to
focus on the "how" questions of the study:

> How does Lucy endeavor to transform information into useful knowl-
> edge?
> How does Lucy's memory work contribute to her status and professional
> identity within the organization?

As I noted at the beginning of this chapter, Lucy spends a great deal
of time and expends a considerable amount of effort in memory practices
intended to help her stay organized and on task. The observation sessions
above confirm this by showing Lucy utilizing an elaborate, redundant, and
multisensory set of practices to ensure that she is able to find the information she needs when she needs it. These practices form part of Lucy's response to an environment of constant change at Software Unlimited and to
her own complex personal circumstances: (1) Lucy's reminding practices are
intended to ensure that her attention remains focused on the correct task at
the correct time; (2) Lucy's archiving practices are intended to acquire and
preserve the history and rationale behind each stage of her product's evolution; and (3) via repetition, Lucy's finding practices are intended to transform information about her product into knowledge. Considered together,
Lucy's practices of reminding, archiving, and finding roughly duplicate the
stages of the process of learning that ancient memory theory made the basis
for classical Greek and Roman education: the learner must first pay attention to the matter at hand (precept), the learner next attempts to preserve or
store this matter by putting it in a form that is meaningful and accessible
(imitation), and, finally, the learner transforms the information into knowledge via repetition (habit) (Murphy, 2001). Lucy's case thus helps us answer
the "how" questions of the study—How does the contemporary knowledge
worker strive to turn information into knowledge within his or her memory

regime?—with the answer that successful knowledge workers come to practice an art of memory appropriate for their embodied contexts.

Many of the specific tools and practices that Lucy employs in this endeavor will probably be familiar to anyone who has experience working in
a contemporary office (e.g., maintaining a schedule on a group calendar,
organizing files on a hard drive, keeping a journal of meeting notes, tracking project deliverables using project-planning software), but the manner
in which she deploys and constellates these practices serves to improve
her ability to achieve both immediate task-related (kairotic) goals and her
longer-term career objectives (phronesis), including, especially, her desire
for colleagues on the software development team to recognize that she is
competent and productive and works as hard as they do. This goal echoes
Lave and Wenger's (1991, p. 173) assertion, stated earlier, that "the primary
motivation for learning involves participating in authentic activities that
move one towards becoming more central to a community of practice,"
and, to that end, her practices demonstrate Lucy striving at every opportunity and via every tool at her disposal to render her information about
her organization and its products comprehensible and knowable. As we go
through this analysis of Lucy's preferred memory practices, I will endeavor
to critique her level of efficiency and success in using them to achieve her
workplace goals.

Before proceeding, however, I need to make an important point about
the level of expertise, or, in our Aristotelian framework, the specific level
of mastery that Lucy aspires to in her job function and performance at this
time. In the foregoing, I described Lucy's longer term "career objectives"—
which may be summed in brief as being determined to show her colleagues
on the software development team (and no doubt, the information development team as well, since both of these teams represent Lucy's communities of practice) that she is just as competent, productive, and hard working
as they are—as a desire for something on the order of a hexis leading to
phronesis. However, the research data presented throughout this chapter
demonstrate that Lucy, like Robert, finds the achievement of knowledge on
a phronetic level difficult.

The reasons for this may be many, but, in general, there are clear indications that Lucy is at least somewhat overwhelmed by the information
overload that she inherently faces in her position, which, as we will see,
causes her to be at pains to document virtually every bit of information
that comes into her workspace. There is even greater evidence that she
feels a tremendous amount of pressure and anxiety over what she seems

to feel is the enormous responsibility to try to learn, know, and keep track
of the history of both the organization and her particular product line—the
"reasonings" that she records in her handwritten notebook, for example,
which we will also see later. To be fair, there is little doubt that Lucy's feelings of being overwhelmed are exacerbated by her awareness of her atypical
work schedule—and concerns about how that arrangement is perceived by
others at the company—and by her considerable responsibilities and personal pressures at home. Moreover, Lucy had returned to full time status
only a month or so prior to the start of this study, and had been assigned
the added responsibility of "documentation lead" for the company's most
complex product.

Given these facts, it is quite understandable for Lucy—or anyone for
that matter—to feel a certain amount of extra anxiety and more pressure to
perform, so this is not a criticism of Lucy in any way. And yet in substance,
Lucy seems throughout all of my observation sessions to have her hands full
just trying to keep abreast of the constant influx of new information while
also attempting to keep a kind of accurate "historical record" of, in essence,
the way "things have always been done" at the company, in part, so that
she might also understand more about the way things should continue to
be done in the future.

What all of this does seem to indicate is that, while it is certainly clear
that Lucy desires to achieve phronesis in terms of an embodied mastery or
proficiency over the basic aspects of her product line specifically and, more
broadly, over her job functions and performance in general, achieving such
a phronesis may be a little beyond her grasp at this rather critical time in
her career path, as well as in her particular circumstances at the time of this
study. That is, it seems fairly evident that Lucy has enough on her plate in
terms of achieving the base level of techne as a competent and recognized
member of her team, and making a definable contribution to her communities of practice—the things she affirms that she presently strives for. We
will see further evidence to support this interpretation and assessment as
we proceed through this discussion.

LUCY'S REMINDING PRACTICES FOCUS HER ATTENTION

Over the past year, Lucy has had to adjust to managing a growing family,
coping with her child's diagnosis, returning to full-time work, and finding
herself as documentation lead on a new work team. Combined, these adjustments have left Lucy feeling disoriented and struggling to stay on top
of her job duties: "With kids and stuff I get really scattered." Significantly,

when articulating what she considers to be the primary obstacle to focusing attention, Lucy employs a metaphor from family life: "Once I'm started on something, I'm good. But lots of times, if I'm not getting started on something I kind of can't concentrate. It's like you go and find a cup in the living room, and you take it in the kitchen, but on the way you see something in the dining room, and then you start cleaning that . . ." As this metaphor so elegantly reveals, Lucy perceives her principal concern in managing information to arise not so much from the volume of incoming information (i.e., from the high number of issues clamoring for her attention) as from her need to prioritize these concerns so that she can direct her attention to the correct issue at the correct time. This is principally an issue of kairos: Lucy needs to know the appropriate place and time and object on which to focus her efforts. Once she determines these, she is fine (e.g., "Once I'm started on something I'm good").

Lucy's choice of words is very telling here, because in singular fashion, this last statement directly affirms the Aristotelian conception of phronesis—of knowledge as something we are, or as something we do. Once Lucy is "in the flow," like a ballet dancer on stage, or the pilot orchestrating the cockpit controls, she is "good." In speaking about tacit knowledge, which we may equate to some degree with phronesis knowledge, Spender (1996) states that "tacit . . . must be an appeal to a form of knowledge with which we are all intimately familiar, the kind of knowledge we pick up by 'osmosis' when we join a new organization or take up a new activity, and on which our sense of domain mastery is based" (n.p.). Referencing Goodenough (1971), Spender continues: "Goodenough . . . defined culture as what one has to know in order to be taken by the natives as one of their own" (n.p.).

Goodenough's somewhat colloquial phrasing neatly and succinctly describes Lucy's present "career objectives." But we can take Spender's pronouncement further: "The notion of culture as *confident activity* draws attention to praxis rather than to *abstract theorizing*" (Spender, 1996; n.p.; emphasis mine). On the one hand, Lucy's sense of "flow" or "confident activity" and the degree to which she finally begins to feel comfortable in her role as an information developer within her community indicate her desire to achieve bodily hexis above all else, at least as things stand at present. But on the other hand, recalling that phronetic knowledge, for Saugstad (2002, p. 383) "is defined as intuitive expertise . . . [and] an intuitive ability to estimate a given situation's possibilities in relation to . . . general knowledge, rules, and principles"—that is, the capacity for abstract thinking— Spender's assertion provides a further indication that Lucy's efforts have so far fallen short of achieving phronesis. Lucy is acculturating in both body and mind to being back in the office full time, but she is not there quite yet.

With that being said, in order to stay organized and focused—to regather her scattered attention, as it were—Lucy employs the most thorough set of reminding practices of any of the information developers. Unlike Robert, who largely dismisses the importance of reminding practices to his ability to perform his duties, Lucy remains quite cognizant of the importance of these practices to her ability to stay focused and organized. Whereas the forward momentum of the Agile development cycle often causes Robert to overlook the importance of reminding practices, this momentum has nearly the opposite impact on Lucy: it makes her only too conscious of her need to do everything she can to stay on task.

Further complicating Lucy's situation, her staggered work schedule makes her keenly aware that she needs to be (and, equally critically, to *appear* to others to be) as productive as possible during regular working hours so that she can be maximally productive and so that her coworkers will know that she can accomplish as much as they can: "I try to make sure that I can get the most out of the time that I am here [and] that I am able to keep everything going and show the progress on it." Lucy, in other words, recognizes that the constraints on her time and attention will make it difficult for her to prove her worth to this new community to which she has been assigned.

Lucy's desire to squeeze every productive moment out of each working day has led her to implement a particularly rigorous and thorough set of reminding practices to keep her days organized and her mind focused. To achieve organization and focus, Lucy believes she needs first to achieve a high-level perspective: "I like to work from the big picture." The big picture will, Lucy believes, make transparent the network of dependences and interdependences that influence her tasks and those of her teammates. This perspective will, in turn, help her concentrate and stay focused on the most timely and appropriate (i.e., kairotic) task. So, for example, when discussing a Sprint planning meeting, Lucy notes her frustration at her product team's method of identifying project deliverables: "We have a spreadsheet and we project the spreadsheet, and as we come up with the tasks, I type them in. And I really pushed them to put it on the whiteboard instead . . . because it was really hard to see exactly how many things, the relationship between those things, and [to] really define priorities when you are scrolling up and down."

Dissimilar to Robert, who prefers richly detailed descriptions over high-level summaries on whiteboards, Lucy strives to achieve a holistic view of her projects that will enable her to perceive the network of relationships and interdependences within the project at a single glance. Rather like a memory theater or memory palace in the rhetorical *ars memoria*, the board also provides affordances that enable Lucy to perceive relationships between tasks given physical, metaphorical representations, so that she can perceive the hierarchies and relationships among the items she needs to remember and prioritize. In terms of the ancient mnemonic advice of Cicero, Lucy's memory *imagines* (images) need *loci* (places); as Small (1997, p. 7) notes, "Cicero more directly says that 'an object cannot be understood without a place." Some memory advice remains perennially relevant: A spreadsheet displayed on a small laptop screen is simply unable to simultaneously display all project tasks of all team members and the relationships of those tasks. A spreadsheet does not let Lucy see the big picture in a manner that would make the information memorable.

Lucy believes that her product team frequently runs into trouble as a consequence of this inability to see the big picture. Despite the emphasis in the Agile methodology on transparency and empirical feedback from all team members, which is supposed to be achieved via the Daily Scrum meetings, neither Lucy nor her teammates appear able to adequately perceive the network of task dependences: "I'm having [problems] because things aren't getting done. Because I don't know when things should be done. . . . But sometimes something is supposed to be done and no one has even mentioned that they've worked on it or are going to work on it or anything because we are missing something with this tracking." This has led Lucy to experimentally adopt another reminding tool, the web-based project-planning software 37signals' Basecamp—in order to help her see these dependences and accurately track her own deliverables: "I thought I'd try to use Basecamp . . . and I tried to put those milestones in. I tried to do to-do lists to go with those milestones for me. I can't say that I'm using it perfectly, but at least I have this to-do list of all the things that I have to do and what they are dependent on." However, Lucy has so far been unable to convince the other members of her product team to utilize Basecamp, so the tool has proven to be of limited use to her.

In marked contrast with Robert, who evinces a strong dislike for Microsoft Outlook, Lucy maintains her efficiency principally by utilizing the reminding affordances offered by the Outlook calendar. Lucy essentially schedules every working hour of every working day in a shared Outlook calendar so that both she and her teammates are reminded of the work she accomplishes and the meetings she attends. Lucy uses her Outlook calendar schedule to plan each workday down to the hour, even going so far as to create entries for her blocks of "free time," which she can use for extended composing tasks like authoring user guides: "I try to track what I'm doing every day—the whole day—in my calendar. . . . Usually I try to make sure I have the next day planned before I leave. So I put all these things in there. Like here's an hour for this—and I already have all my meetings in there so I put them [the hours of free time] between my meetings." However, from experience, Lucy knows that colleagues who see "free time" in her shared calendar may receive the false impression that she has too little to do or, more immediately detrimental, may schedule meetings during these precious spans of time she uses for composing. So, to ensure that she has the extended time she needs for composing and writing, Lucy flags these blocks of time "private" so that her colleagues cannot see that they are open. "I do better when I can concentrate on things for a long time rather than doing a little bit here and there. . . . That's really what my schedule helps me with." Lucy's many competing obligations mean that her attention is at a premium. The Outlook calendar helps manage her attention by limiting potential sources of distraction.

At the end of the week, Lucy also uses her Outlook calendar to remind both herself and her colleagues of her progress during the week: "It helps to jog my memory for what I really did and to be able to show it." In other words, Lucy attempts to make her Outlook calendar entries play multiple roles: the entries are memory tools with both reminding and archiving affordances (i.e., they provide visible and audible alerts of upcoming tasks or events and they live on in the calendar database), but they also make Lucy's otherwise invisible work visible and quantifiable. That is, because her colleagues on her product team may otherwise only notice that she leaves early each day, not that she also comes in early and works from home when needed, Lucy needs some way to demonstrate the actual hours she puts into her job. Her Outlook calendar provides this record, and in Lucy's view, it proves for her product teammates that she works as hard as they do and that she is capable of participating as a full member of their community of practice, even though she follows an atypical schedule. It is an open question, however, whether anyone else in that community of practice ever consults that record: in the first place, given the hectic and fast-paced work ethos at the company, it seems highly unlikely that any of Lucy's coworkers would have the time to do so (except, perhaps, Lucy's manager, Becky, when performance review time comes around). More important, though, is the question of what, if any, constructive reason any of her peers could possibly have—and I stress *constructive* reason—for looking at Lucy's calendar record. The prevailing impression one obtains from Lucy's compulsive hours-on-the-job record keeping is that it serves primarily to bolster her own feelings of participation and work effort.

With this in mind, it remains to be seen whether Lucy's Outlook calendar entries actually enable her to simultaneously get her work done more efficiently so much as they enable her to preserve a tangible history proving that she works equally long hours. Primarily, it seems that by enabling Lucy to view all of her weekly accomplishments at a glance, the Outlook calendar helps her feel better about herself and the job she does, giving her a psychological boost: "Now I feel productive. I may not be and no one else may have noticed, but I feel more productive." This statement is instructive in that it indicates fairly plainly that this issue for Lucy, whether she fully realizes it or not, is largely one of self-perception, and acknowledges that, at some times at least, she may not have had a productive week. While her Outlook calendar schedule may be said to contribute both to Lucy's perceptions of her ability to meet the kairos of her daily work situations and to the cultivation of an embodied hexis—her own and her colleagues' perception that she is a competent, organized, and efficient professional, and thus a master of her working environment—it does not seem to confirm that she is meeting the criteria of kairos and hexis *in fact*. On the positive side, Outlook does appear to serve both prospective and retrospective functions for Lucy in helping her to organize her work and prioritize her tasks, which after all is the main purpose of a calendar/scheduling tool in the first place.

However, the work sessions presented above illustrate yet another function that Lucy appears to rely on through the use of the Outlook calendar: Lucy uses her calendar entries, in perhaps something of a default manner, to support task performance. This somewhat unorthodox use of the calendar can be seen in the first session. Early in that observation, Lucy appeared to be unsure of the precise version of the document draft she wanted to work on during the work session: She examined several potential files in Windows Explorer, but browsing in this way did not provide her with enough information to make a decision. To recover from this lapse, Lucy opened her Outlook calendar entry for the currently scheduled event where she had copied the file path and name of the document she planned to work on during that period of time (see fig. 5.1). By copying and pasting the file path as if it were a website URL, she was able to quickly find the file. A positive interpretation of this "recovery" might be that, because she had the forethought to append this extra bit of information to her schedule entry, Lucy was able to find the correct version of the file and quickly resume work. And in fact, when I questioned Lucy about this practice in a short interview following the session, she noted that she often appends this type of functional information to her Outlook calendar entries. In other words, the practice apparently forms a routine part of Lucy's memory work; on the surface, it appears to be an innovative and clever, albeit extra, step she takes when putting together her daily schedule that ensures that she is able—ultimately—to locate the information she needs when she needs it during each work session. But is it efficient, and, moreover, is it indeed an instance of achieving "mastery" over her work environment?

In fact, it would appear not. Lucy did not use this method to find the file right from the start. Instead, she began her search by browsing her Sprint folder. Here I could speculate that Lucy may have been exhibiting an innate human preference for browsing, were it not for at least two important factors. First, throughout both observation sessions we see a characteristic in which Lucy indeed wastes a lot of time browsing and searching what appear to be redundant files, even opening multiple files at a time, looking for either the "latest" version or, alternatively, the one she needs or has scheduled herself to work on. After spending significant minutes searching files and not finding the one she wants, Lucy resorts to searches of file folders and help files, only to have those tactics fail, whereupon she returns to browsing. At one point, she generates a new help file that she subsequently recognizes has something wrong with it. All of this activity seems rather haphazard, hit or miss, and decidedly inefficient, and one is left to wonder why Lucy doesn't go directly to her calendar first to avoid all of this wasted time. If, as noted earlier, Lucy is neither intimidated nor terribly concerned about the amount of incoming information she has to deal with, it appears that her method of dealing with that information overload—by apparently creating numerous redundant versions (or multiple versions each with minor daily updates, thus confounding the finding of the "very latest" version)—is not optimal by any means. The result, as we have seen in both observation sessions, is that Lucy unwittingly creates unnecessary confusion for herself both in identifying and finding the right files (i.e., most up-to-date versions) that she needs. Thus, as comfortable, secure, and "productive" as this overdocumentation makes her feel, it ultimately causes her to waste valuable time which in turn makes her less efficient and less productive.

Second, while the preference for browsing might still be at play subconsciously, such a casual approach to file searching simply does not accord with Lucy's burning desire to show herself as competent, efficient, and as productive as her team members within her communities of practice, which would seem to argue for a much more pinpoint, instantaneous, and reliable method. Further, Lucy's fallback reliance on her Outlook calendar—a tool more specifically designed for schedule tracking—to look for searchable information that is really more appropriate to one's document database may be very telling. Specifically, reliance on her calendar is another indicator that Lucy may be somewhat *over*concerned about the perception of her work performance among her colleagues at the company, making her overprotective of her self-image (as a result of the pressure to perform that she feels) to the point that her efforts to keep track of everything and "show progress on it" are actually negatively affecting her ability to do just that.

Still, there is no mistaking through all of this that Lucy is striving mightily to perform in a manner that qualifies her as someone "central to a community of practice," to use Lave and Wenger's phrase. And even if, by itself, Lucy's practice of appending extra information to her calendar entries does not appear to be a major innovation, or a time-saver at all, it is nevertheless typical of the ways in which Lucy jigs her environment to facilitate her tasks and, therefore, demonstrates the sort of situated expertise that Star (1999, p. 387) labels "street smarts." Star defines street smarts as the "real-time adjustments, or *articulation* work" of the savvy and experienced worker that "weav[es] together desktop resources, organizational routines, [and] running memory of complicated task queues" to overcome disruptions and achieve maximum efficiency (p. 387). Lucy's dual-purpose use of her calendar as both a reminding and an archiving memory practice may also be an indication of her effort, in Clancey's (1997, p. 4) words, to coordinate "activity within activity itself." The practice serves as evidence that, via her extensive experience working for companies large and small, Lucy has learned to take advantage of as many affordances as her tools offer her: her calendar not only reminds her of upcoming appointments and preserves a record of past accomplishments; it also enables her to make "real-time adjustments" with minimal disruption to her ongoing tasks (Star, 1999, p. 387). I would suggest, in other words, that practices like this are evidence that Lucy is striving to achieve a hexis, or the kind of mastery of the memory-enhancing affordances of her workspace that will enable her—on a fairly basic and embodied level—to overcome routine disruptions without losing focus or attention, even if those efforts are not entirely successful.

LUCY'S ARCHIVING PRACTICES PRESERVE THE HISTORY AND RATIONALE BEHIND HER PRODUCT

Lucy believes that her archiving practices are just as important to her success as her reminding practices. The principal motive for Lucy's archiving practices arises from her belief that she must master what she terms the

"reasonings"—the historical events and political circumstances—lying behind design decisions affecting the company's products. What should be
immediately apparent in this belief is the strong influence of Angela, whose
storytelling has had a profound and powerful impact on Lucy. As a result
of her interaction with Angela, Lucy is principally guided by the characteristics of the memory regime within her community of practice, shaped by
Angela, which accords privileged status to all such historical knowledge
about the evolution of the company's products and the difficulty that non-
software-developers have in obtaining and understanding this knowledge.
These characteristics of the regime compel Lucy to become preoccupied
with amassing as much historical information in as rich detail as possible.

Another motivating factor influencing Lucy's archiving practices arises
from her extensive previous experience working for a much larger company.
Before joining Software Unlimited, Lucy worked for five years as part of a
team of nearly one hundred technical communicators. Because of the disparity in the size of Software Unlimited as compared to her earlier firm,
Lucy has had to learn to adjust to a more informal, self-reliant approach to
managing information. Even though it has been two years since she worked
for this larger company, Lucy emphasizes that the drastic change in culture
continues to influence her attitudes and approaches toward memory work.
For instance, where at her former company, she could be assured that her
daily work would be preserved via automatic backups, now Lucy must be
consciously proactive in her archiving: "We had very specific things that we
did. We had to keep things. Our computers were backed up every night. We
didn't have to do anything. I didn't have to do anything then." Now, Lucy
must remember to save and back up her files to the network drive each eve-
ning. In a sense, this situation puts the onus on each individual information
developer to independently amass and preserve all of that historical information, the "reasonings" that Lucy has come to regard as so vitally important.

As with Robert, Lucy's initial inexperience with the memory regime at
Software Unlimited led her to make false assumptions about how the company did things when she first joined: "Coming from the background I did,
I would expect certain things and then I'd do something." Without a great
deal of more formal, company-based archival "doctrine" of the sort that
might be "backed up automatically" every day and preserved by the firm
itself, it should come as no surprise that Lucy would inevitably turn to her
more senior information developer colleague Angela for an explanation of
what she had done wrong and how (and why) to do them the preferred way:
"Angela could explain to me the three-year reason—you know, the reasoning over the last three years—of how something had gotten [the way it is]."

It is worth pointing out here that there are indeed circumstances under which storytelling practices such as Angela's can work very well. It is quite easy to imagine a situation in which the newer information developers at the company are collectively unclear about many of the same company procedures or processes that they all must somehow deal with, whereupon the cry goes up to "ask Angela," as will be shown in chapter 6. As a smaller organization (as the company was two years earlier, back when Lucy was hired), it would have been a simple matter for Angela to call an impromptu meeting of her teams, at which she could recite the narrative history of the "world according to Angela." In fact, when Angela first joined the company, when it had only twenty employees, an even more pervasive and effective storytelling strategy would have been to get all twenty employees packed into one room in order to recite the narrative history in full for everyone to absorb and accept, and to become, in effect, "indoctrinated"; doing so regularly would ensure that everyone, including the inevitable newcomers to the growing organization, would "keep the faith." With that picture in mind, it is quite easy to understand why Lucy is so profoundly influenced by Angela directly, and by the historical stories ("reasonings") emanating from Angela's preferred memory regime.

Thus, because of Angela's advice, Lucy believes that she has been able to achieve a better understanding of the memory regime of Software Unlimited, and, consequently, from the very beginning, she began keeping a detailed archival history of the decision-making processes lying behind each stage of the evolution of whichever of Software Unlimited's products she worked on: "There was a lot of stuff like that around here—you know just being a small company with all this stuff: it was like all these reasonings. So I think that's why I started collecting a lot of that, the reasoning part, because it makes a difference around here." In fact, Lucy, like Robert, has learned through experience that historical knowledge about the company's products is at least as important as understanding a product's current feature set or future design trajectory: "People expect that you know the reasoning, like why it is the way it is." Lucy preserves this historical information principally via her careful archiving practices. We might speculate that when Lucy says, "People expect that you know," she is really saying that Angela expects her to know. But let's examine what Lucy actually does.

First, Lucy creates and maintains a file folder database of product information organized by Sprint. Adhering to the classification conventions advocated (but not mandated) by the organizational memory regime, Lucy organizes this database into folders and subfolders by Sprint number and by product feature name (the feature name refers to a piece of product function-

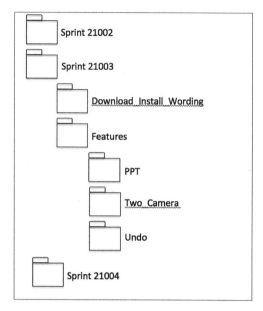

Figure 5.4   Lucy's file folder database organized in one-month increments by
Sprint. She names the subfolders in each Sprint folder for the functionality
or features that she is working on during that release. Lucy places her
major deliverables and supporting documents in these folders.

ality that her Scrum team works on as part of each thirty-day Sprint goal
(see fig. 5.4). Lucy indicates that the visual hierarchy of this database (i.e.,
the folder, subfolder, and file icons viewable using Windows Explorer) plays
an important role in helping her stay organized and keep track of project
information: "So there's not a list of all the documents and all the features,
just different folders because then I also put in there anything [related to
the release]." That is, the file and folder structure itself serves as a kind
of visual organizer helping Lucy see at a glance the scope of each Sprint in
comparison to other Sprints.

Within this file and folder structure, Lucy stores all the documents
and files related to each stage in the evolution of her product. The most
important documents within this database are what Lucy refers to as her
"data dump" documents. In these data dump documents, Lucy attempts
to preserve as much detail about each release of her software product as
she possibly can: "What I try to do is capture what has gone on during the
month we are working on something so that the next time we work on
it I know where the product is. [Including] everything I know about this
feature that we are doing right now with screenshots of where it is." Al-

though these data dump documents resemble early versions of the software manuals that she writes for major releases of her product, Lucy creates them more for archiving purposes than to give herself a head start on writing the documentation: "It's nothing I could put out—I don't even consider it like a draft. . . . I actually document the process of how it works right now knowing that that isn't going to be the final process." Creating the data dump documents represents, in other words, a memory practice distinct from writing documentation.

The data dump documents enable Lucy to respond to both short- and long-term concerns. First, in terms of kairos, having already authored multiple drafts of the documentation during the design cycle helps Lucy create final drafts swiftly during the crunch time at the end of a Sprint: "They won't have to go through a lot of revisions." Similarly, if she finds that she needs to create additional documentation or deliverables during a Sprint, the preliminary drafts give Lucy a resource from which to draw: "I might have to dual purpose them—that'll be a standalone piece that we can use for tutorial or some kind of learning piece and then I'm also going to integrate that information."

Nevertheless, Lucy acknowledges that creating these extensive and detailed data dump documents probably represents extra work on her part, work that will not necessarily play any part in her final deliverables: "There's probably way more information in it than I'll ever put out." Some additional explanation appears necessary, then, in order to account for why a professional with such a keen awareness of the demands placed on her time and attention undertakes this extra, nonessential work—work, moreover, that remains largely invisible to her colleagues on her development team.[1] Lucy indicates that this motivation is related to her efforts to achieve hexis: writing helps Lucy learn and master her product. In other words, the act of repeatedly documenting the varied and often transitory evolutionary steps in the design of her product creates both external *and* internal representations of the product. Whether this process, in the future, ultimately will make Lucy more proficient, more productive, and more valuable in the eyes of her colleagues remains to be seem. However, the process effectively contributes to the problem of redundancy that I alluded to earlier in this chapter, and, as we have already seen, it presently has a negative impact in causing Lucy to at times waste valuable time searching for the "latest" version of documents.

Externally, of course, creating these data dump documents provides Lucy with a usable (i.e., searchable and browsable) external archive of historical details about the evolution of her product, albeit one that appears on the basis of the observation sessions to be somewhat unwieldy. Still, this archive can serve as a resource for both Lucy and any information developer colleague who might need to fill Lucy's shoes: "Like if something happened and I couldn't finish it up, somebody could see where the product had gone or where that feature we had been working on. And with how much stuff we've got going on and how long some of these things take, I don't always remember and have to look back and see." At the surface, all of this intensive record keeping in her effort to in essence "archive everything," including successively older (and thus successively more out-of-date) versions of documents, may reflect Lucy's compulsion (and her nervousness) to preserve a viable archive or "backup" of her own, in response to the fact that Software Unlimited, the company, does not perform an automatic, company-wide, fail-safe backup of its own, the way her previous employer routinely did. More importantly, though, the act of writing helps Lucy internalize historical knowledge (i.e., the "reasonings") about her product's evolution: "Each Sprint we were working on stuff. So I just started 'boom' here's what was going on with this [feature] at that time with a screenshot. . . . I have a tendency to, like, I absorb things and then I just write." Consequently, the process of composing the final versions of the documentation becomes easy: "[Because] I produce that stuff I think that the actual writing part, like when I'm actually sitting down to write the text, is the easiest part and kind of the no-brainer part of my job. . . . By that time, I know the product, what's actually going on, and I know the history of how we got there, which I think is huge because collecting stuff as we go along." In short, Lucy ascribes her success in one of her most important job duties—that of authoring documentation—largely to her practice of keeping meticulous archives. Archiving practices enable Lucy to meet the kairos of the multiple daily work situations of her techne, but they also contribute to her hexis, her longer-term ethos with her teammates by teaching her about the product.

A third archiving practice that Lucy engages in is keeping a "documentation library plan" for each new project to which she is assigned. This plan, composed as a large Word document, serves a variety of audiences and purposes, both memory and communication related: "I go through and each feature I have a table that says the name of the feature [in] the heading. . . . That's kind of the first part—describing the feature with enough information so that the developer could know that we know what this feature is going to be." Lucy brought the idea of the library plan with her from her previous job and adapted it for her needs at Software Unlimited. Lucy's adapted library plan has proven so effective at preserving and communicating project information that it has been adopted by the rest of the information development team and, indeed, has become the standard template for maintaining the memory of a project over the long term. In this, the team's acceptance and adoption of Lucy's library plan in large part both demonstrates the esteem in which her information-managing skills are held at present and serves as an example of the process by which an individual's innovative memory practice can come to influence the memory regime of the entire organization—or, at the very least, that of the individual's more immediate community of practice.

Most importantly from Lucy's point of view, however, as with her other data dump, the library plan enables her to preserve the kind of background or historical context for design decisions that she regards as crucial to success at Software Unlimited. Echoing the language she uses when describing historical information preserved in her data dump documents, Lucy notes that the library plan enables her to preserve "the whole reasoning behind [each feature], so that I could go through it later and I could see that my first initial instincts were for how this feature was going to affect [the documentation]." Significantly, one result of the diffusion of the documentation library plan to the other information developers is that Lucy's awareness of the importance of preserving background knowledge has spread to the rest of the team. As an example of this awareness, the team manager, Becky, indicated that the library plans had become an important vehicle for preserving historical information: "If the next writer picked [a library plan] up, they'd have a history. All of them do it a bit differently but the idea is to sort of present to the team 'Here's what I think we are going to have to do to support this product.'" (It is interesting to note that Becky, who as the information development team manager is Angela's "boss"—appears to have also bought into the importance, most powerfully espoused by Angela, of "preserving historical information.")

A final archiving practice—and another example of the diffusion of idiosyncratic individual practices throughout an organization—is Lucy's keeping of a project notebook, which she carries with her to all product team meetings. Lucy adopted this practice after observing the representative from the training team at previous product team meetings: "He always goes around with this little . . . notebook . . . and this thing of pens and they are all different colors. And he always takes this everywhere, and what he does is, every meeting he writes down [meeting information]. [Then] he prints out meeting notes, meeting requests, and everything—real small—and he glues them into his book. . . . He's always able to go back and find stuff." By adopting and modifying this practice to suit her own needs, Lucy endeavors to consign—that is, to render findable—any paper record related to her team meetings via her project notebook: "Every meeting I have is in here. Actually, I have it color coded between [the software product team, the information development team] and the training team."

Lucy believes that this new archiving practice compares favorably with her earlier, more reminding-like practice of taking notes at meetings: "much more than I ever used any of those notebooks that I just, you know, had randomly." As with her other archiving practices, Lucy envisions one of the principal benefits of this memory practice lying in its effects on her ethos with her product team: "It's been very helpful. . . . And I think it helps me be a lot more productive. . . . I can go back and see if we made decisions or whatever. . . . I know what's going on and I know that I know what's going on. So I speak up much more often than I used to on a lot of that stuff. . . . They [i.e., her product team members] want me there." By including additional contextual information concerning team decision-making processes in her notebook, Lucy creates, she believes, a transportable archive that ensures that those all-important "reasonings" remain always at her fingertips. The notebook, therefore, plays a key role in helping Lucy speak and act as a full member of her software development team. Where the software developers had previously neglected to include her in meetings, she is now, thanks to this practice, considered a valuable asset to the community. The developers make sure they invite her to meetings because they want her to share her insights and knowledge. Lucy's notebook, therefore, becomes a tool of both individual and community memory, enabling and encouraging Lucy to talk as a full member, but also helping the group preserve and use its own memories. The notebook both "support[s] communal forms of memory and reflection [and] signal[s] membership" (Lave & Wenger, 1991, p. 109). More, perhaps, than any of her other archiving practices, then, Lucy's notebook assists her in achieving phronesis.

As with her reminding practices, Lucy's archiving practices display a frequently overstretched but always-savvy knowledge worker employing every tool at her disposal to render her complex information environment manageable. These practices are, in other words, a sign of Lucy's ongoing (but not entirely, as yet, successful) efforts to achieve expertise on the order of phronesis, and, as Kirsh (1995, p. 36) notes, "The hallmark of expertise . . . is sufficient compiled knowledge to cope with normal contingencies without *much* on-line planning [and] a major factor in this compilation is expert perception: having the right perceptual categories, and knowing how to keep an eye on salient properties." Lucy's archiving practices are good indicators of her attempt to achieve this "expert perception."

LUCY'S FINDING PRACTICES TRANSFORM INFORMATION
INTO KNOWLEDGE

Merely possessing or storing up information in archives, even when those archives are easily portable (such as Lucy's notebook), is not the same as learning, however. The memory practice of archiving is only one step in the transformation process by which information is turned into knowledge. To acquire the kind of phronesis, the state of mastery that the ancient rhetoricians, Bourdieu, and situated-cognition theory all identify as the ultimate goal of a practice, the learner must in some way absorb the information at the level of embodied dispositions (i.e., hexis). The ancients, Bourdieu, and the situated cognitivists also all agree that the chief mechanism for this absorption is repetition leading to habit. As Carruthers (1990, p. 68) states, "The organism's hexis or habitus is developed by the repetition of particular emotional responses or acts performed in the past and remembered, which then predispose it to the same response in the future. . . . Experience is made from many repeated memories, which in turn are permanent vestiges." I suggest that close analysis of Lucy's finding practices, particularly her preferred practice for refinding information stored in her archive, reveals that the act of finding is itself a key mechanism through which Lucy strives to achieve mastery of product knowledge.

The majority of Lucy's memory practices in the two observation sessions presented earlier are finding practices, whether specifically attempts to locate new information or to relocate information that she has previously encountered and archived. Lucy's preferences when undertaking these finding practices follow a distinct pattern, which can be summarized as follows: when attempting to find information, Lucy appears to prefer location-based browsing over text-based searching *except* when she perceives time to be a major factor, such as when a progress-stopping breakdown occurs or when the sheer volume of information appears to be overwhelming. Lucy begins every instance of routine information retrieval in both observation sessions by browsing: In order to locate all three documents in the first session and her working document in the second, Lucy first navigates through her folder structure in Windows Explorer (see, e.g., fig. 5.2), and, when attempting to locate information online on the WebWorks website, she begins by browsing the site's table of contents before using the site's search feature (fig. 5.3). She only employs text-based searching as a finding strategy in instances where browsing appears not to be an option, such as when she attempts to locate a specific piece of text in a lengthy Word document, or as a last resort after browsing fails to return the results she wants, such as when she abandons the WebWorks wiki to search the online help.

While this preference for browsing is hardly unique to Lucy and has been widely observed in other studies of information-seeking behavior, browsing is inherently less efficient and can be both less accurate and more likely to mislead the user than searching (e.g., Barreau & Nardi, 1995; Marchionini, 1995; Jones & Teevan, 2007). As Marchionini points out, users are more likely to be "distracted, confused, disoriented, or frustrated by peripheral and tangential information" when browsing than when searching (1995, p. 118). Marchionini asserts that browsing may even be a prime cause of the sense of information overload that information seekers sometimes experience. Therefore, he notes, "the value of the information seeker's time and any time-sensitive charges may . . . discourage browsing" (1995, p. 118). Also, browsing via the desktop file folder metaphor, such as Lucy strongly prefers when attempting to locate information in her desktop archive, has come in for particular criticism for insufficiently supporting multitasking and for confusing information access and display (e.g., Kaptelinin & Czerwinski, 2007). Consequently, browsing represents nonoptimal behavior: the more experienced and familiar with an archive or other information space that a user is, the more likely he or she is to employ search to save time and effort (Ravasio & Tscherter, 2007).

These limitations of browsing do appear to negatively affect Lucy's performance in both of the observation sessions, as noted earlier. Indeed, browsing appears to contribute to the four interruptions in Lucy's work processes observable in these sessions: Lucy experiences delays finding (via browsing) each of the three documents she works on in the first session and is altogether unsuccessful in locating the first document she looks for in the second session (she ends up finding a similar document, renaming it and using Find and Replace to change its content). Further, Lucy's behavior indicates that she remains cognizant, at least to some limited extent, of the relative inefficiency of browsing in comparison to searching: instead of scrolling down and scanning her long Word document, she does not hesitate to use the Find feature of Word to jump to the section she is looking for; she similarly shows no patience for browsing the WebWorks online help but instead moves instantly to the search feature (a not uncommon behavior owing to the perceived difficulty of finding information in online help systems: e.g., Krull et al., 2001; Grayling, 1998; Pratt, 1998). Yet none of this stops Lucy from automatically starting virtually all of her searches by browsing first.

So why, given the number of search tools available to her, the comparative speed and accuracy of such tools, and the tendency for browsing to lead her astray, does Lucy persist in browsing? As we have already noted, this preference seems especially out of place for someone who, like Lucy, employs all of her other memory practices to maximize her efficiency and to focus her attention, and who, as well, is extremely concerned with demonstrating her efficiency and productivity to other members of the company, and with doing so in the most visible ways possible. The research on information seeking identifies several advantages of browsing that, given both what we know about Lucy and what we have learned about her extreme desire to maximize her productivity over time, may help to explain her persistence in this otherwise nonoptimal practice.

First, Marchionini (1995, p. 102) notes that browsing provides information seekers with an overview of a "physical or conceptual space" with which they are not familiar. We browse, in other words, to "discover and learn" (Marchionini, 1995, p. 105). To illustrate this, Marchionini offers the example of a scholar who, upon encountering a new book in her field, might scan the title page, table of contents, section headings, and index in order to get a sense for what the book is about and whether or not it will be useful (p. 102). Jones and Teevan (2007, p. 26) echo Marchionini when they speculate that browsing helps users understand how a piece of information relates to other pieces of information with similar characteristics because it "can provide both an overview of the information space . . . as well as context about where the desired information is located." This use of browsing would probably help fulfill Lucy's desire to achieve a panoramic view of a problem space before attempting to work within it: "I like to kind of work from the big picture." A glimpse of the way browsing helps Lucy see the big picture can be seen in the second session when she carefully examines the table of contents and submenu on the WebWorks website (see fig. 5.3). Browsing the WebWorks site lets Lucy know all of the user assistance resources that are available to her before she commits time to viewing a single one.

Second, Marchionini postulates that browsing can help information seekers monitor the status of an ongoing process. This function can be particularly helpful, Marchionini adds, where "the objective is to stay abreast of a field" (1995, p. 103). For example, a scholar attempting to keep up to date with breakthroughs in her discipline might visually scan a list of abstracts in order to spot recent additions. An indication that Lucy relies on this affordance of browsing can be seen in her carefully labeled and chronologically ordered archive of Sprint folders: merely by scanning down the list of subfolders, Lucy can quickly perceive if an aspect or feature of a given Sprint release is missing or out of order (see fig. 5.4). In this way, browsing enables the archive to serve a sort of reminder-like function for Lucy. This may be especially helpful in the fast-paced Agile development environment at Software Unlimited. In other words, by expanding the Sprint folders to display the features as she has done in figure 5.4, Lucy can ensure that no feature gets overlooked and forgotten from one Sprint to the next.

Marchionini observes that a third appeal of browsing probably lies in the cognitive assistance it offers to users. According to Marchionini, browsing assists the information seeker's memory because the environment itself displays some of the data needed to complete the information-seeking task. In other words, rather than having to completely mentally disengage from a task in order to, for instance, brainstorm key words or search terms, a user navigating a browsable archive is able to seek information visually, which does not require her full attention or concentration.[2] As I noted earlier, maintaining focus and attention is one of Lucy's primary goals when undertaking her various memory practices, so, in this regard, her preference for browsing makes sense: browsing does not compel her to interrupt her task completely in order to shift into finding mode.

Other research suggests that a fourth advantage of browsing is that it gives users a greater sense of control over their information. According to Barreau and Nardi (1995, p. 40), users appear to prefer browsing because it "imparts a greater sense of control" by "more actively engag[ing] the mind and body" in information-managing tasks. In other words, interacting with digital representations of information or memories, such as with the files and folders on their computer desktops, makes users feel as if they can touch and manipulate their information to suit their needs and purposes.

This explanation for the appeal of browsing accords well with what we know about the information developers' relationship to the memories of Software Unlimited. As the analysis of Robert's practices in chapter 4 demonstrated, direct knowledge of the company's software products is the most important indicator of expertise at Software Unlimited, but the only community members who are able to make direct contact with these products (i.e., who are able to touch the software code enabling them to see the stages of a product's evolution in the code's revision history and in the comments of previous programmers) are the software engineers, not the information developers.

It seems likely, then, that the sense of control over information that browsing imparts plays a significant role in Lucy's preference. This supposition is supported by the fact that all but one of the other information developers indicated a similar preference for browsing. For example, Robert's principal archiving practice, his "back of the mind" folder, as discussed in chapter 4, represents precisely the kind of carefully organized filing structure that encourages browsing. Similarly, Monica, in answer to my interview question asking if she found that the process of managing projects grew more difficult as the project continued, replied that it did not, thanks to the careful file folder structure and naming conventions she employs at the beginning—also habits that facilitate browsing: "I feel like I do a pretty good job at the beginning trying to set things up—a structure. We did a really good job, I think, of organizing our N drive, and I try to follow that and put documents where they should go. I have multiple files or folders and piles of paper. I usually get the initial structure for how I want to set up the documentation, where I'm going to put [it] on the N drive." When asked a similar question about how he got up to speed when he first joined the company, Peter couched his answer in terms of browsing too: "I'd find out that there's a folder on the N drive called 'current software'. And then once I did go there to look for it, I looked at everything else that was there, too. So I did a little exploration as I needed to."

Becky, for her part, characterized her success at managing multiple projects simultaneously as a result of the carefully organized folder structure she maintains in Outlook: "As I start getting stuff, I usually for every project have a specific email folder so I can file everything. And if it gets too big I break it down, like 'meeting notes' and different pieces of what I have to do. And in my in-box if there's anything I have to address, it doesn't go in a folder so I'll know I haven't addressed it yet." In fact, of the information developers, and perhaps not surprisingly, only Angela indicated that she did not value browsing or utilize the kinds of visual metaphors that facilitate browsing, like email in-box folders and elaborate file folder structures. Her explanation for this preference is also explainable in relation to the issue of control, a topic I take up in chapter 6.

To summarize, current research reveals that browsing offers at least four potential benefits as a finding practice: it can offer a big-picture view of an archive or information space, it can help users monitor ongoing processes, it can help preserve users' attention, and, most significantly for the information developers at Software Unlimited, it can impart to users a sense of control over information. However, this research appears to point to a further benefit of browsing—more specifically of what I will refer to as "habitual browsing"—in that repeated, habitual browsing of the same information over and over, while appearing on the surface to be redundant, nevertheless may serve a useful function in helping to "cement" or solidify information or knowledge firmly into one's memory, and with greater "permanence" over time than the sort of cursory browsing that mostly amounts to little more than a scanning of headlines or article titles, for example. This may be a relatively new phenomenon originating in the new digital technology which allows one to browse, essentially at one time, a greater volume of information much more quickly than one could before the existence of computer screens and virtual pages, and perhaps an area deserving of future study. But with respect to this study, habitual browsing does appear to help Lucy achieve sense of control over information, as well as possibly a sense of hexis over her job function.

In short, I am arguing that Lucy's browsing exerts a strong mnemonic effect: the activity of repeatedly navigating up and down her hierarchically organized archive of product information (i.e., the files and folders) enables her to learn and master the history and evolution of her product. This activity suggests that what Brown, Collins, and Duguid (1989, p. 37) assert is largely true: "The structure of cognition is widely distributed across the environment, both social and physical. And we suggest that the environment, therefore, contributes importantly to indexical representations people form in activity." Although Lucy, like the other information developers, is provided with access to her product's interface after each build, this view of the evolving interface is partial and limited, only giving her a view of the product at a given moment.[3] In the absence of access to the source code, browsing a file folder structure organized by Sprint and feature is, therefore, as close as Lucy can get to achieving a holistic view of the evolutionary stages of her product. Perhaps most importantly for the mnemonic effect, however, because it entails manipulating representations of her product (i.e., by clicking, double-clicking, dragging, and dropping file folder icons within the database), browsing habituates Lucy to both *seeing* and *interacting with* representations of her product's history. As Brown, Collins, and Duguid (1989, p. 37) further state, "Indexical representations developed through engagement in a task may greatly increase the efficiency with which subsequent tasks can be done." In other words, this habit is formed at the level of mind and body. Clancey (1997) characterizes this nicely within the methodology of situated cognition as follows: "The theory of situated cognition . . . claims that every human thought and action is adapted to the environment, that is, *situated*, because what people *perceive*, how they *conceive of their activity*, and what they *physically do* develop together" (p. 2; emphasis in original). Viewed in this way, browsing is one of the methods through which Lucy converts information into knowledge, and it may exemplify what Clancey (1997) sees as a "shift in perspective from knowledge as *stored artifact* to knowledge as *constructed capability-in-action*" (p. 4). Lucy's goal, in the context of "constructed capability-in-action," is to achieve a virtuoso command of her knowledge systems and their contents and represents her drive to achieve hexis leading to phronesis.

This embodied mnemonic aspect of browsing can be perceived by comparing Lucy's verbal account of her database with the account offered by her use of that same database in the work sessions. Lucy describes the file folder archive of product information that she keeps on the hard drive of her computer, as follows:[4]

> I have a Sprint folder [tapping the table with the index finger of her right hand: number 1 in fig. 5.5], and then in there I have different things. But I have one folder called "Features" [pulling her hand toward herself and to her right along the tabletop: number 2 in fig. 5.5], and in the folder "Features" I have a different folder for each of the things that we are working on [continuing the direction of the gesture: number 3 in fig. 5.5]: "Power . . ." [interrupting herself], "PPT" [pulling her hand straight toward herself: number 4 in fig. 5.5], um "Undo," um . . . [looking off into the distance for a couple of seconds before pulling her hand toward herself: number 5 in fig. 5.5] "Camera Recording." Just everything that we are working on and then I have those documents in there.

Lucy experiences two memory "lapses" when offering this description. First, Lucy starts to describe one feature folder as "PowerPoint" but interrupts herself and changes to the folder name, "PPT." Second, in a nearly perfect example of a movement that language and gesture researchers refer to as a "speech failure gesture," Lucy pauses and looks off into the distance while trying to recall the name of another subfolder, which she ends up calling "Camera Recording" (McNeill, 1992, p. 77).[5] Lucy is right-handed and uses her right hand exclusively to control her computer mouse when working. Yet, with the exception of the imagistic gestures in which she describes using her computer mouse, such as those occurring in the transcript above, Lucy tended to be ambidextrous when gesticulating.

The form of these gestures, therefore, appears significant from a memory perspective in that it seems that the memory that fuels Lucy's description of her product information archive is contained not just in her mental image of her folder structure but also in the kinesthetic memory of her right hand. That is, the motion of Lucy's hand as she describes the folder structure in the interview mirrors the motion of her hand as she guides her cursor to navigate the folder and subfolders of her data dump in Windows Explorer. Lucy's hand follows the same path. In short, Lucy's gestures are,

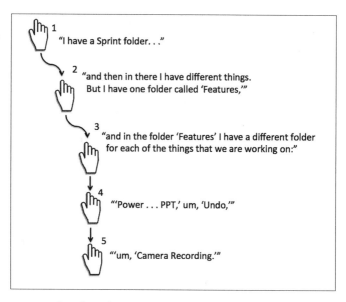

Figure 5.5   Lucy describing the contents of a Sprint folder. As she describes each of the
contents, she gestures by sliding her right hand along the tabletop.

as McNeill (1992) predicts, revealing her "memory image" and "particular point of view" of one of her typical activities, but they are also showing that this memory image is composed of both visual and kinesthetic components (p. 13).

Equally significant, comparing Lucy's gestures with the momentary lapses noted in Lucy's description of her archive—her midword self-correction from "Power-" to "PPT" and her pause to remember another subfolder related to the camera functions of her product—it appears that the gestures are not merely helping her *display* her memory image of the folder structure, but are actually helping her to *remember* the information, to actively construct it from memory. That is, if her attention was focused primarily on describing the folder's contents with maximum clarity, Lucy would probably have finished the word "PowerPoint" and left it at that, but, as her pauses and frequent "ums" indicate, she instead appeared to be focusing a great deal (perhaps the majority) of her attention on just remembering those contents. So, as she scans her mental image with the eye of her mind and touches it with the memory of her hand, she lists the contents as she sees them—"PPT" instead of "PowerPoint," as figure 5.5 illustrates. As McNeill puts it, "Gesture can materialize the spatial structure of the problem. . . . Spatial structure is a mnemonic, especially when external-

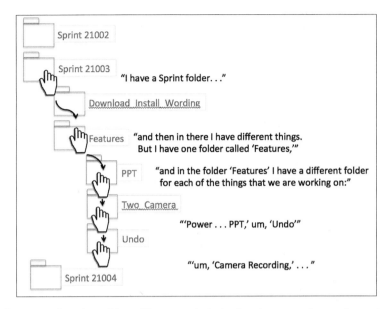

Figure 5.6 Superimposition of figure 5.5, depicting Lucy's memory image of a portion of her Sprint database as revealed in her gestures as she describes this database, on figure 5.4, depicting the contents of one of her recent Sprint folders. The result illustrates the mnemonic effects of habitual browsing: the memory image, an Aristotelian phantasm, adheres in hand and mind; kinesthetic memory reinforcing semantic memory.

ized" (2005, p. 148). By superimposing Lucy's folder structure on her mental image of her folder structure (see fig. 5.6), it becomes apparent that Lucy's picturable and touchable folder spaces serve a multisensory mnemonic function for her as she attempts to remember and articulate their contents. Lucy's database illustrates Aristotle's phantasm at work.

This incident highlights, then, the other major function that browsing serves for Lucy: it is a practice that translates product knowledge, the knowledge most highly valued in the memory regime of Software Unlimited, into a form that she can see and interact with. In particular, Lucy's file folder hierarchy inscribes the relevant aspects of her product's evolving interface and functions into a structure that she can touch and manipulate. By repeatedly browsing this structure as she accesses and uses her working files on a daily basis, Lucy memorizes the product's history and trajectory, thereby completing a learning process initiated each morning when she arrives at work.

In other words, even when browsing fails as a finding strategy, Lucy's finding practices, like all of her memory practices, still perform an important function for her by helping her cement knowledge of her product. That is, in the same way that we become familiar with a route by repeatedly walking through that landscape, Lucy's activity of browsing her file folder structure implants knowledge about her product firmly in her memory. These practices, then, contribute to Lucy's acquisition of a hexis.

CONCLUSION

As I discussed in chapter 1, memory researchers from a variety of scientific and academic fields have in recent years converged on the idea that embodiment is critical to effective memory work: we best retain information perceived via multiple senses, and we are best able to recollect and use this information when given the freedom and the tools to interact with it via multiple modalities. In particular, situated-learning theory distinguishes between raw information and useful knowledge by reference to embodied reality: knowledge is information that has been processed; knowledge "usually entails a knower" (Brown & Duguid, 2002, p. 119). Thus, situated-learning theory posits that the activities and contexts in which information is encountered and used are essential to our ability to retain the information as knowledge. Perceptions at the time of use then become keys to using the information: "Knowledge . . . indexes the situation in which it arises and is used. The embedding circumstances efficiently provide essential parts of its structure and meaning. . . . Knowledge . . . comes coded by and connected to the activity and environment in which it is developed" (Brown, Collins, & Duguid, 1989, p. 36). Knowledge, in other words, is information that has somehow been brought *into* the knower via multiple "sensory gateways" (Carruthers, 1990, p. 12). Once the information has been perceived and noticed, we typically think of it as residing in the memory of the knower; it is now possessed in some sense by the knower. The knower remembers to one degree or another the experience of encountering the information, and the information thus becomes usable for future purposes: it can be refound when needed.

The most obvious manifestations of the types of embodied knowledge identified by situated-cognition theory are found in manual trades or in the performing arts, where bodily skill and habit can easily be seen in the measured precision of the master craftsperson or in the choreographed footwork of the trained dancer. We tend not to think of such physiological training as a craftsperson or dancer undergoes as playing a significant role in the practices of office workers like Lucy. And yet, Lucy's case illustrates that something very similar to the sensorimotor training of a skilled performer takes place every time Lucy browses the archive she has so carefully assembled: her repetitive actions are bringing the outside in.

Considered together, then, Lucy's memory practices of reminding, archiving, and finding recapitulate a learning process that, according to Aristotle, leads to hexis, "a state . . . that follows on perceiving, apprehending, experiencing, or learning" (Sorabji, 1972, p. 1). Moreover, as Carruthers (1990, p. 68) points out, such embodied training leads to the kind of assured practical skill married to social understanding that constitutes phronesis: "The basic connection between the process of sensation which ends in memory, and that of human emotional life is fundamental for understanding the crucial role memory was thought to have in the shaping of moral judgment and excellence of character." Lucy's memory practices are, as Robert's obsessive archiving practices are to him, the primary means through which she integrates herself into the community of practice of her software team, but they are also, concurrently, signs of her efforts to achieve phronesis.

The research questions that this chapter engaged with entailed answering "How" questions: How do technical communicators transform information into useful knowledge, and how does this memory work contribute to the status and professional identity of technical communicators within their organizations? The answer we can offer to these questions is that the information developers both transform information into knowledge and use this knowledge to gain status in their communities of practice and the organization as a whole by enlisting multiple embodied senses in the tasks of memory work. Lucy's preference for browsing over finding is shaped by a combination of her own individual needs and the cues and constraints offered by the memory regime, but a common motivation underlying both Lucy's and the other information developers' memory practices appears to arise from their desire to experience and engage with information in multimodal, multisensory ways in order to render it more memorable: more easily found, more easily learned, and more easily recollectable and recollectable.

As I noted above, Lucy, like Robert, spends a great deal of time and expends a considerable amount of effort doing memory work. Lucy utilizes an elaborate, often redundant, and frequently multisensory set of cues to ensure that she is able to find the information she needs when she needs it. That is, in order to manage in an environment in which information from multiple spheres of life is continually clamoring for her attention, Lucy has learned to take advantage of whatever affordances she perceives in her surrounding physical, digital, and social environment. In short, Lucy is continually learning and adapting in order to stay abreast of constant changes in work processes and projects at Software Unlimited while simultaneously managing the daily exigencies of her complex work and life circumstances. Lucy, therefore, strives at every opportunity and via every tool at her disposal to render her organization's history and her own work schedule visible and tangible and, hence, rememberable.

CHAPTER SIX

Mastering Memory

Telling a story in which you are a character yourself creates constraints as well: You become a character with a specified role in the subsequent stories of the listener/reader and you cannot permit yourself too much deviation from the expectations connected with this role.
—J. Jasper Deuten & Ari Rip, 2000, pp. 80–81

The focus of this chapter is Angela, whose nearly six-year tenure with the firm makes her by far the most senior member of the information development team. This comparatively lengthy tenure gives Angela a unique historical perspective on Software Unlimited and its products and both its internal and external processes that, theoretically at least, ought to make her an invaluable resource for the other information developers: she knows the stories behind the products' evolution, and she knows how "things work" at the company. Analyzing Angela's memory work, therefore, presents an opportunity to consider issues of information and knowledge that affect full members of communities of practice and to consider the roles that narratives and stories play in circulating memory among teammates. In particular, it also allows us to consider the condition of phronesis within a memory regime, and to examine how that mastery-level expertise flows to, and is emulated (or modified or transformed) by, the other members of this community of practice. As Brown and Duguid (2002, p. 107) state, "The value of stories . . . lies not just in their telling, but in their retelling. Stories pass on to newcomers what old timers already know. Stories are thus central to learning and education." Further, Angela's case encourages us to consider why and in what ways possessing the stories and history that she possesses helps her to work more effectively, as well as why and in what ways reliance on those stories and history may actually hinder her effectiveness. As the analysis will show, phronesis, once achieved, is not a permanent condition.

Before coming to Software Unlimited, Angela had already experienced a long and varied career in industry. In the early 1990s, before becoming a technical communicator, Angela had worked in retail management. In order to advance her management career, she had started taking college courses in the evenings, where a perceptive writing professor quickly recognized her talent:

> After class one day the instructor said "Can you stay? I want to talk to you about a paper you've written." And I hadn't been to college in a long time, so I was like "Oh no, what is that about?" But she literally said, "I just want to know what you are doing here. You should be a writer or a communicator, and if you decide to go to college full time and take this on, then I think that you and I would be a good fit together to come up with something—a new type of degree to fit these computer things that are coming out."

Together, Angela and the professor put together this "groundbreaking and innovative" degree, so that she was able to double major in technical communication and English. After graduation, Angela held several full-time and contract positions related to technical communication. During this time she cultivated a specialization as an expert in process improvement and worked with several major corporations.

Angela was not initially hired by Software Unlimited to be a technical writer per se but was recruited by the founder and CEO to be a process improvement manager. At the time, Software Unlimited had fewer than twenty employees, and the CEO was attempting to move the company's two products from a shareware to a commercial model. That is, the products, which had been free (with correspondingly low expectations for functionality), were being rebuilt so that users could be charged for them, which would raise user expectations that the product function well and provide adequate instructions. Consequently, the CEO needed Angela's expertise to "for the first time ever, put in workflows and processes into the systems." Angela's expertise proved instrumental in helping the products evolve over the next several releases. She was, in other words, instrumental in guiding the evolution of the company's original products into the form they exist in today.

Gradually, as the products succeeded and Software Unlimited grew, Angela's job role became more formalized as the company's first information developer, which, as I noted in chapter 3, entailed user advocacy during design as one of its principal responsibilities. The origins of the information developer's role in steering the development of the user workflows and interface, therefore, can be traced directly back to Angela's own specialization as both technical writer and process improvement manager. And Angela had a hand in recruiting each of the other members of the information development team, including the second-most-senior writer, Lucy, who had been hired "around two years ago," and the team manager, Becky.[1] While Angela's history at Software Unlimited places her in a unique position at the firm, many aspects of her career path will be familiar to her generation of technical communicators, including her unusual degree in an era before many undergraduate degree programs in technical and professional communication existed (e.g., Davis, 2001), her experience as both a consultant and full-time employee of companies large and small (e.g., Barker & Poe, 2002), and a career path ranging from solo writer to member of a team of technical communicators (Kim & Tolley, 2004). Angela's example has much to offer us in understanding the roles that memory practices play in the work of senior members of work teams.

Because of Angela's extended tenure with the firm, the other members of the information development team understandably regard her as a full member of their community of practice, but, more than this, they also regard her as a master of institutional memory, a crucial resource for the community. As evidence of this, the information developers repeatedly reference Angela's thoroughgoing knowledge of Software Unlimited's history and how important this history is to acting successfully within the organization. The information development team manager, Becky, for instance, notes of Angela: "She's got the most history of anyone in our group. . . . We are really lucky because she's got so much background and understands sort of how we got to where we were with some of these choices and different things that were tried. So we are really lucky to have that. Some of the teams don't have as much history." Peter similarly asserted, "Anything about why things are done a certain way she would know because she's been here the longest." But it is Lucy who, having been mentored by Angela the longest, most highly values Angela's long-term perspective and command of the organization's stories. In fact, Lucy expresses admiration bordering on awe of Angela's prodigious institutional memory: "A big thing that I've noticed here is all of the reasons—back reasons—Angela's been huge. I can't even imagine going into that mind for all the stuff she has in there. . . . Angela could explain to me the three-year reason—you know, the reasoning over the last three years of how something had gotten [the way it is]." By most indications, Angela is highly valued by her colleagues on the information development team; her command of the institutional memory appears to make her an invaluable ally in the effort to warrant arguments with the development teams, and even to some extent the marketing team. Lucy appears to have benefited the most from these stories.

The information developers are each indicating in their own ways that they believe that Angela has achieved the sort of situated expertise, or phronesis, to which they aspire: she is not only an "old-timer" in their community of practice; she appears to them also and more importantly to be a master of the organization's memories. They believe that Angela has achieved this state because she has acquired the narratives or stories of the organization and its products and that her retelling of these stories provides them with crucial context and the background rationale that has driven the design decisions affecting all of the firm's products.

As shown in chapter 4, one of the primary reasons that the information development team considers Angela's mastery of this context and background information so important probably stems from the fact that such historical information is not readily accessible to those who are not software engineers. This is not to imply that the memory regime at Software Unlimited encourages any sort of deliberate or malicious withholding of critical information, but, rather, it is a recognition that some types of knowledge are inherently more difficult to preserve and transmit than others. In other words, while the company's archives may contain project-planning and implementation documents detailing "explicit" knowledge about company projects and products, these official accounts, like similar accounts in any organization, often do a poor job of capturing "tacit" knowledge of the politics, the rhetoric, or the processes of deliberation surrounding the explicit knowledge contained in official documents (DeLong, 2004, p. 101). This was one of the lessons of chapters 4 and 5: newcomers and employees other than software engineers experience difficulty finding the information they need to do their jobs over both the short and long term. Consequently, Robert and Lucy must exert extra effort to preserve as much contextual information as possible about the company's products and their development trajectories. As we saw in those chapters, they apply different sets of memory practices in their efforts to accomplish this task.

Storytelling thus appears to be the primary memory practice engaged in by Angela as her means of sharing knowledge with her information developer teammates, and such stories are excellent vehicles for conveying precisely the kind of background context and deliberative detail that official archival genres like project planning documents often neglect: "Stories can be effective for transferring both implicit knowledge about how things get done, as well as deeper tacit knowledge that reflects the values shaping behaviors" (DeLong, 2004, p. 102). Or, as Brown & Duguid (2002, p. 106) put it, stories not only explain *what* or *how*; they also explain the *whys* crucial to the achievement of phronesis: "Stories are a powerful means to understand what happened (the sequence of events) and why (the causes and effects of those events)." By virtue of her tenure with the firm, as well as her long working relationship with the founder of the company, which will be discussed below, Angela's stories represent one of the information development team's most important sources of knowledge.

Such contextual background knowledge is, or at minimum has been, as chapters 4 and 5 also demonstrate, especially crucial to the ability to work effectively in the particular organizational memory regime of Software Unlimited. There is no doubt that Angela believes this is still true, and her teammates seem generally to concur, though Lucy appears to do so to a greater degree than does Robert. Lucy attributes the importance of knowing the stories to the small size of the company and notes that Angela's command of the stories motivates her own habit of meticulous memory practices: "There was a lot of stuff like that around here—you know, just being a small company with all this stuff: it was like all these reasonings. So I think that's why I even started collecting a lot of that, like, the reasoning part—because it makes a difference around here." However, the question that is begged, as the company undergoes rapid growth and departmentalization, is whether this is still true. For example, Lucy continues by speculating that she thinks this aspect of the memory regime of the company may be changing because of the company's recent growth: "I don't know if we are kind of getting away from that [the importance of knowing the reasonings] as we get a bit bigger."

The growth of the company is already having an impact on the role that Angela's stories play within the organization. This impact can especially be seen in Angela's interactions with the community of practice centered on her product team. By her own account, Angela emphasizes that she believes she has a very strong relationship with her product team: "Our team is probably one of the closer teams here. We spend a lot of time trying to build that relationship with each other and do a lot of team building events and things, which isn't typical across development teams." Angela believes (quite naturally, given her seniority) that these teammates regard her as a full member of their community of practice. In fact, Angela has worked alongside some of these teammates on the same software product for six years: "From the day that I got here I was always part of [software] development and I sat with the development team in their area." As this statement indicates, Angela's office is even located with her product team rather than with the other information developers, who all sit together in an information development area (see fig. A.1 in app. A). This proximity enables Angela to glean information from her product team far more easily and effortlessly than either Robert or Lucy: "So on a continuous basis, if I run across something in the product as I'm using it, I can get up out of my chair and holler around the door: 'Do you have a minute? Can you come and look at this crazy thing that just happened?'"

However, it would seem that this proximity also serves to reinforce Angela's preferential reliance on storytelling as her primary memory practice. In essence, it seems most likely that knowledge that is obtained or transformed out of informal information-sharing by "hollering around the door" is not likely to be recorded or input to other, more formal—and more sharable—memory systems, such as the company's computer database systems. What seems much more likely is that Angela will simply take knowledge gained from such informal conversational episodes and add them to the overall narrative of her storytelling memory regime.

Not only that, but there also already appear to be cracks forming in this seemingly transparent and "smooth" cross-community knowledge-sharing relationship (as Angela sees it). The company's recent rapid growth and consequent Agile-driven reorganization into cross-functional product teams have resulted in a product team drastically different in composition and expertise from the one Angela was used to. Many of the new team members, including representatives from the new departments of user experience (UX), training, and marketing (which are also emerging communities of practice in their own right), do not know Angela's history with the product, nor do they understand her expertise in user testing workflows and interfaces. Indeed, one may go so far as to suggest that team members in these newly formed departments do not "respect" or recognize either Angela's history or her self-proclaimed expertise, specifically, because functions such as user testing and marketing are tasks that *they were hired to do independently* (i.e., not by taking their direction from Angela). In other words, the new people in UX, for example, do not see Angela as part of their community of practice in the same way that Angela does.

Understandably, all of this makes Angela uneasy: "As we grow, things get more complicated." Compounding her unease, Angela's product team is also working on the biggest new release of her product in several years—since, in fact, the first commercial release in which Angela implemented the original workflows. This new version will change those original workflows and interface in radical ways, and Angela's efforts to guide this rapid evolution of her product—and in a very real sense to retain control of it—forms the backdrop for our investigation of Angela's memory work discussed in the following sections of this chapter, and for the subsequent analysis.

Before turning to this review and analysis of Angela's memory work and practices, one other aspect of her workplace identity warrants consideration. To almost the same degree that they emphasize her command of the history of the organization, her fellow information developers also emphasize Angela's creative bent, particularly her visual creativity. According to Peter, "She's great with graphics and anything Graphic Forge related . . . and she does stuff with it that I just never thought of before."[2] As her manager, Becky particularly appreciates Angela's creative savvy and penchant for experimenting with novel and innovative tools and approaches: "I think Angela's got a lot more tools; she's got a lot of tricks up her sleeve; if you need things, she's got a whole array." Angela, in particular, believes creativity to be a distinguishing feature of her professional identity, one that sets her apart from the other information developers.

But there is an ominous side to this perception. Consider Angela's words: "There are many people on my team that are . . . interested in the technical communication society and memberships and content management and structure. I'm more on the 'let's be creative side.'" If Angela is indicating that she sees herself as somehow different from the run-of-the-mill technical communicator concerned with standard technical communication issues, and further, if she takes pride in being spontaneous, imaginative, and innovative, she might, charitably, be allowed some of that distinction based on her position as the senior member of the team, and the sheer fact of having the longest "war record" of any of the information developers. But what this self-perception also suggests (which represents a preview into what we will learn about her memory practice preferences later in the chapter) is that Angela relies very heavily—in fact, almost exclusively—on her knowledge of the historical narrative of the organization in general, and on storytelling as a memory practice in particular. As a revealing example, in addition to differentiating her from more mundane matters of technical communication, Angela believes that her creative side influences her research and information management habits. For instance, when discussing her research habits, Angela, again distinguishing her behavior from that of her information developer teammates, emphasizes that "a lot of people read a lot of blogs and online journals about communication. I tend to go to digital scrapbook pages and the online artwork pages to get new and innovative ideas for the way people are using digital media to tell stories."

In the course of this chapter I will explore the impact that Angela's reliance on storytelling has, both positive and negative, on both her performance and her perception, her "stature" among her colleagues on the information development team, as well as those on the software development
and product teams. But suffice it to say for now, one has to be concerned
about these remarks by Angela. While her less experienced information
development team members are reading up on the latest trends, developments, and innovations in the field of technical communication, Angela
is browsing pages devoted to artwork and interesting and trendy ways that
people are telling stories with digital media. At a minimum, Angela is missing out on potentially important and valuable information, which is her
responsibility to assimilate, manage, and turn into knowledge that will be
helpful to benefit the company. At the same time, however, it is also clear
that Angela's creativity does contribute to her ethos with the information
development team, and this fact in turn relates positively to her mastery in
the community of practice. I would therefore venture to say that Angela's
mastery, in the eyes of her information developer teammates, approaches
or attains phronesis; however, it is not at all clear that it does so among
members of the other communities of practice with which she interacts or
to which she belongs.

FIRST OBSERVATION SESSION: ARTICULATING EXPERTISE

During our initial interview, I asked Angela, as I asked each of the information developers, what software tools she used in her composing activities
at Software Unlimited.[3] In contrast to the answers to this question given by
the other information developers, which were short and "stubbornly 'factual'" (Gabriel, 2000, p. 26) lists of authoring tools and their purposes, Angela answered with an extended historical narrative:

> One of the goals that has been in place since I started with Software Un-
> limited was design a Graphic Forge product that "Angela can actually
> use." When I got here and started documenting and looking at Graphic
> Forge, there were a lot of things that were not in Graphic Forge so I
> could use it to do all of my screen captures. It was good at taking screen
> captures, but once I got the screen capture, there were different things
> that were in the compression algorithm and whatnot that weren't right
> for what I needed. And the workflows may have been there for a lot of
> people, but they weren't right for technical writers.
>   So we came up with the add-ins at that point, and we developed

those. So rather than having to take me out of my environment to have to go to Graphic Forge to do the capture and then do something with it then save it and come back to Word and import it as a file, we developed—conceived of and developed—these add-ins where you're right inside of Word and you can click the Graphic Forge button and go take it and do whatever with it and save it and then actually link it back as a file all in one fell swoop.

So that was phase 2 of "if Angela can use it." And then we came up with the output from right within Graphic Forge. So now we had backward compatibility backward and forward, but this will actually be the first release in which we are using Graphic Forge to develop our graphics—no other third-party software. A lot of places rely on Photoshop to put finesse on their graphics or to save them in a certain way. So it has truly been a goal for the past five years to make Graphic Forge as robust as possible so we can use it. So now if you look at our help files for this release, you'll see little tags in the corner that say "created with Graphic Forge." And part of actually being able to use Graphic Forge to create our help files, part of that is a sales pitch to people, that hopefully they'll be able to open up the help file, to look and see what we have done. And it will serve as examples and self-proclaiming tutorials on how they can do the same kind of work. So we are hoping to directly impact the conversion ratio through all of that.

A couple years ago I put out a series of creative tutorials which taught people how to do Photoshop-like interesting graphics using Graphic Forge. So if you see the photo cube and different perspective and shared graphics on the edges and things like "how to do, how to do portraits" and "how to take a graphic and make it look like a hand-painted portrait." All those things that you really need a lot of specialized knowledge in Photoshop, I taught people how to do it in Graphic Forge. Those were some of the best tutorials we ever did, because it got people really excited about using the product and how to look like a superstar using just Graphic Forge. So this is just taking that idea that I had to [the] extreme case of actually using it to create the files or the graphics that go in. And they're not just screen captures at this point. The whole help is graphic versus just text and a graphic.

So if you download, say, the OneNote add-in for Graphic Forge, the OneNote help actually fits the personality of the OneNote product itself. So the OneNote help when you download it you'll see instantly that it is completely different than any help file you've seen. Rather than a graphic and some text in a file, it's got a collage/scrapbook kind of look to it. It has a background of notebook paper being torn out and inkblots here and there like a grocery list. Like OneNote is supposed to be used as a notebook.

So we started with those help files and the community at large started noticing that our help files were being transformed. So I started carrying that through to our ancillary documents like Graphic Forge notes or the blogging tools and things like that. So now all the online help and all the ancillary documents are being updated for this new release. So this is going to be the biggest concerted effort I've ever made of using Graphic Forge, being creative, using my collage format, and putting it all together.

SECOND OBSERVATION SESSION: DEFENDING EXPERTISE

Throughout the early days of my on-site research, Angela appears especially enthusiastic about the focus of my study and expresses great enthusiasm for sharing information about her memory practices. About a month into this research, she invites me to observe and study her weekly meeting with her counterpart from the training team, Carl, which is to take place later that afternoon. Overhearing this, Becky adds that Angela's memory practices will be particularly enlightening because of her long history with the company.

Angela informs me that she has called this meeting with Carl in order to discuss the number and content of the training tutorial videos needed to support the upcoming major release of their product, Graphic Forge 5. Before the meeting, Angela prints two copies of the agenda, one for herself and one for me. We carry these to the meeting. Each member of the information development team has recently been partnered with the member of the training team assigned to his or her product team in order to coordinate the training and documentation efforts, to ensure consistency across documentation, and to avoid unnecessary duplication of effort on each product release.

The meeting between Angela and Carl takes place around a small table in the "Petri Dish" area of the Information Development Wing that is reserved for members of the Graphic Forge team (see fig. 6.1). When we arrive, Carl is already waiting and has set up his laptop on one of the mobile tables in the Petri Dish. He positions his laptop so that Angela can see the screen as they talk. The screen displays a "mind map" that he has created using Mindjet's Mind Manager software. The training team regularly uses Mind Manager to collectively brainstorm ideas, but the information development

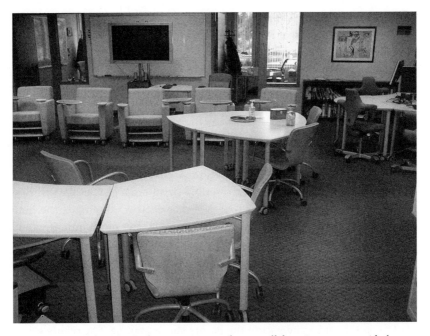

Figure 6.1   The Petri Dish, an experimental team collaboration area set aside for
use by the Graphic Forge team.

team does not use it or have it installed on its computers. Carl's mind map
shows his assigned tasks and his ideas for a series of training videos that
he plans to create for the Graphic Forge 5 release. He also opens the most
recent build of Graphic Forge 5 and deploys the most recent version of the
graphics-based online help that Angela is creating to accompany this major
release. Angela has brought only her printout of the agenda and a pen. She
places the agenda on the table in front of her, but I note with some interest
(and curiosity) that she only doodles on it during the meeting—she makes
no substantive notations. Carl has not brought the printout of his copy of
the agenda (which Angela had mailed to him earlier) but has, it appears,
created tasks for himself in his mind map based on the items on the agenda.
    They begin their discussion by considering the number of training vid-
eos that Carl needs to create for this release. Angela has been pressing for
a one-to-one correspondence between the topics in the online help and the
training videos, but Carl is concerned that he will not have time to create
that many videos. He reasons that the thirty topics in the online help will
translate into far too many videos for him to create in the allotted time.
Consequently, Carl is also heavily invested in this meeting in order to de-

termine which user tasks are most important so that he will only be responsible for producing videos for that subset of tasks.

During this discussion, Carl volunteers some constructive criticism of
the radically overhauled graphics-based version of the online help that Angela has proposed for this release (as she reveals in the previous extended
interview story of the first observation session, in order to showcase the
power of the graphical capabilities of Graphic Forge 5, she has redesigned
the online help so that the text of each help topic is embedded in a graphic
rather than saved as a text file). Carl politely notes that the current visual
format looks good but that he believes the online help offers too many options and that these options are presented in an order that users will find
confusing. He further suggests that Angela's decision to embed the help
topic text in image files might not be very usable (at a minimum, it would
prohibit text-based searching). As an alternative, he suggests that Angela
consider embedding images in text files.

As Carl offers his criticism and makes his suggestions, Angela's body
language changes: she sits up straighter and appears less relaxed and more
focused. When she answers Carl's points, she begins to jab her pen toward
his laptop screen to underscore the points she is making. She rebuts Carl's
recommendations by informing him that she has "talked to users" and that
these users "like" the new graphic-based help. Based on this experience, she
says, she knows that the "community at large" (her term in the interview
for existing users with whom she stays in contact via company-sponsored
user forums) is on her side. Recognizing from her tone and body language
that he has, at a minimum, overstepped the purposes of the meeting in offering these critiques, Carl drops the issue, adding that he will defer to her
superior experience and knowledge because he has only been with the company for six months.

During this discussion, Carl's laptop screen has been the primary focus
of attention, as Carl alternates between the online help and his mind map.
As noted above, while responding to Carl's critiques, Angela begins to make
deictic (i.e., pointing) gestures with her pen toward his computer screen, on
which he alternates between views of the online help and views of his mind
map. She becomes particularly emphatic in these pointing gestures when
referencing Carl's mind map, where he lists his tasks for the current release.
This behavior is so pronounced that I make special note of it in my research
journal. It almost appears as if Angela is attempting to puncture some sort
of invisible barrier between herself and Carl's thoughts expressed on his laptop screen—or, perhaps, to almost literally "poke holes" in Carl's argument.

Near the end of the meeting, Angela tersely informs Carl that she does not have the Mind Manager software and asks him to email her a screen capture of his mind map to help her prioritize her own tasks. Angela adds that she will block time for these efforts in Microsoft's Team System software, a team awareness application used by her product team that enables teammates to view each other's schedules in order to coordinate work efforts and arrange meetings. Angela asks Carl if he uses Team System, but he informs her that he does not use it much, again mentioning that he has only been employed at the company for a few months and adding that he is still adapting to the way things are done there. Carl further adds that he had not experienced a similar degree of concern for team awareness in his previous job and, consequently, has been surprised by this aspect of the work environment at Software Unlimited. At this, the meeting concludes and Angela returns to her office, which is adjacent, leaving Carl to pack up his laptop for the return trip to the other side of the building, where the training team is located.

ANGELA'S MEMORY WORK

At this stage in both of the previous two case study chapters, at which I delve into the memory work of Angela's two junior information developer teammates, I introduced the workplace behavior of both Robert and Lucy quite specifically as *activities* that appear to embody the particular memory practices that each of them favors among the six memory practices that I identify in chapter 2. Similarly, in the two sessions with Angela, we observe her engaged in the memory practices of storytelling, referencing, and gesturing, all of which do the important rhetorical work for her of translating the past into the present:

> In the interview segment, Angela responds to my question by *storytelling* an extended narrative in which she and the product she has worked on since arriving at the company figure as the main characters.
>
> In the training team meeting, Angela effectively shuts down criticism of her online help offered by a junior member of her team, Carl, by *referencing* the opinions of the user community of their product.
>
> Also in the training meeting, Angela engages in *gesturing* at the computer screen as an apparent means of demonstrably—and physically— highlighting her countercriticism of Carl's opinions on the issues at hand.

Perhaps the most interesting thing about these observations, however, is what they lack: any indication that Angela engages in the conventional information-managing memory work of archiving, finding, and reminding. For example, when asked in the interview to describe her memory and writing practices, Angela never offered decontextualized lists of "information items," such as Robert's description of his "back of the mind" folder or Lucy's discussion about the contents of her data dump. Instead, she couched her answers in stories from her work history, particularly in stories concerning the various iterations of Graphic Forge that she has participated in. Similarly, in the training meeting with Carl, rather than taking detailed meeting notes, as Lucy might do in her project notebook, Angela merely doodles. Finally, her gesturing practices may be necessitated by the absence of any archived documentation, which she could have printed out and presented at the meeting and thus used to explain her position based on knowledge from either her own community of practice or the memory regime of the organization as a whole. As the following discussion will illustrate, none of these idiosyncratic practices were unusual for Angela, but all are representative of Angela's regular memory practices. Understanding these practices and their atypicality in comparison to the practices of her colleagues will help further explicate the answers to the "why" questions of the study: Why does memory work matter; and why, if at all, does it enhance—or perhaps degrade— the status and prestige of technical communicators in their workplaces?

ANGELA'S ARCHIVE RESIDES IN STORIES

Angela differs considerably from the other information developers in her archiving and reminding practices. Where all of the other information developers emphasize the considerable time and effort they put into their memory work and make frequent references to the digital and nondigital tools to which they delegate much of this memory work, Angela indicates that she has, as a result of her lengthy tenure with the company, virtually transcended the need for external tools or careful memory work: "Other people, they have file folders and they keep track of things. I just don't."

Paradoxically, though not surprisingly given the above statement, Angela's nearly six years with the company have resulted in the smallest, rather than the largest, conventional file folder archive possessed by any of the information developers. Referring to the often-meticulous (even compulsive) memory practices of colleagues like Robert and Lucy, Angela notes, "There are several people on the [information development] team who are very, very structured. [But] I've been doing this so long and I've been here . . . a long time. I'm able to just know a lot of stuff—historical stuff—and why we did things." And, as the interview segment demonstrates, this "historical stuff" is, for Angela, chiefly preserved in stories rather than in external archives and databases.

These stories not only preserve details about the evolution of both the company and her product; they also give form and shape to this information, transforming what would otherwise be disconnected data into genuine knowledge, knowledge that helps Angela understand both what happened during each evolutionary stage and why the evolutionary step occurred in the first place. Stories do this because the "logical and semantic connections" established by their narrative arc (once upon a time, this happened, and this led to that, which led to this other thing) preserves the sequence in which past events occurred (Fentress & Wickham, 1992, p. 50). Knowing the proper sequence, in turn, preserves the cause-and-effect relationship between events: "Stories are good at presenting things sequentially. . . . They are also good for presenting them causally. . . . Thus stories are a powerful means to understand what happened (the sequence of events) and why (the causes and effects of those events)" (Brown & Duguid, 2002, p. 106). In other words, a story possesses powerful affordances for memory other forms of archives or databases lack: "A story is thus a large-scale *aide-mémoire*" (Fentress & Wickham, 1992, p. 50).

Where newer employees like Robert and Lucy are concerned that, to succeed in either the long or the short term, they must vigilantly create and maintain detailed archives ordered by Sprint (see fig. 5.4), Angela's stories provide all the context that she believes she needs in order to do her job. Moreover, where the other information developers must preserve their memories in external archives and are only able to internalize these memories via repetition, such as Lucy's browsing in chapter 5 accomplishes, Angela's stories are already internalized and embodied because they originate from her own firsthand experience. In other words, as her interview responses demonstrate, Angela herself is every bit as much of a character in the story as the product and the process through which the product is developed; Angela's work history and the evolutionary history of her product are intertwined. As her account reveals, in order to advocate for users Angela does not believe she needs to track down the results of various usability tests, nor does she have to attempt to think like a prototypical user, as her information developer colleagues do. Instead, she was, quite literally, the first user of her product and the product has, in her view, evolved to meet her needs ever since. It would be difficult to imagine a closer integration of a worker's infrastructure for doing work with the object of that work than Angela's.

Finally, even in those instances when Angela does engage in archiving, she does so mostly to benefit her newer colleagues on the information development team, rather than because she believes the archive would be a resource that she herself would use:

> Whereas people who are new coming on board, they don't have that historical perspective and so they have a need for that information and so I spend some down time if I have any documenting things about the way they were or how we do things because it used to just be "ask Angela," "ask Angela," "ask Angela." So, I think I try to remember as much as possible, and we have a place where other people can go for stuff, if they can find it.

It appears that Angela resorts to archiving only as a means of reducing the number of times she must repeat the telling of her historical stories in response to all of those "Ask Angela" questions from her less experienced team members. When the team and the firm were smaller, Angela was able to share her knowledge with everyone who needed it via storytelling. Hence Lucy, the second information developer hired by the firm, could benefit from being given the whole backstory: "Angela could explain to me the three-year reason—you know, the reasoning over the last three years of how something had gotten [the way it is]." But, with ever-increasing numbers of new employees clamoring for explanations, Angela no longer has the luxury of telling her stories in full but must instead consign the knowledge they contain to the company's databases. In a sense, it appears in this instance that the memory practice of storytelling breaks down, or even outright fails within the setting of the training meeting (it is notable that Angela does not even attempt to recite for Carl the full story or even part of the story of her decision to create image-based online help for her product, but simply asserts that she knows what her user community wants). For Angela, archiving is, at best, little more than a time-saving substitute for storytelling and, at worst, an effort to reduce the amount of repetitive storytelling she must engage in for the benefit not of herself, but of others in the company.

REFERENCING TEAMMATES MAKES FINDING AND REMINDING EASY

As with archiving, Angela's finding and reminding practices also differed considerably from those of the other information developers. First, as would be expected from someone who disdains structure and who avoids archiving whenever possible, Angela indicates that browsing is not her preferred method of finding information. In fact, Angela disdains the time and effort that would be required by archiving practices to render information browsable in the first place. For instance, of Outlook email, which Becky and Lucy spend so much time organizing, Angela asserts: "Typically email is one of the biggest ways that we communicate, and it's not the best way. . . . Not only just emailing and sending email, but sorting it, filing it, cleaning it up, deleting it, you know all the stuff that comes with email." Consequently, Angela makes no effort to sort and organize her email in-box into separate folders, but, somewhat shockingly, lets it pile up in her in-box: "you know, I've got fifteen hundred emails right now in my in-box, and six hundred of them aren't read yet."

Her disdain for practices of keeping information that would in most of our experiences be critical to our ever being able to find that information again also characterizes Angela's attitude toward reminding practices. For instance, Angela several times alludes to the fact that she does not take notes at meetings, but, as noted in the training meeting, she prefers instead to doodle on a piece of paper: "Doodling frees up some place in my brain that helps me free think on something that's being discussed. In some way doodling helps me focus more on the conversation and think through the complicated processes." Angela similarly avoids that most ubiquitous of reminding practice, creating to-do lists, by asserting that her to-dos reside exclusively "in the back of my head." She believes that she does not need a piece of paper to remind her what she needs to accomplish.

As with archiving, Angela appears to believe that she has, in a sense, transcended the need for reminding practices. This sense of transcendence is equivalent to the state of hexis—the perfectly trained memory condition—to which the other information developers aspire and toward which they hope that their meticulous information-managing practices lead them. Both Angela and her colleagues understand, or believe, her hexis to chiefly result from her lengthy tenure with the firm. In theory, this tenure has made her an eyewitness to the history of the evolution of the firm and its products, thereby giving her the real stories behind these evolutions, and thus the answer to the question of "why things are the way they are" at the company. Equally importantly, Angela believes that this tenure, almost by itself, has also enabled her to develop a particularly close relationship to her product team: "I worked with the developers to, for the first time ever, put in workflows and processes into the systems." These close relationships also, she believes, contribute to her ability to transcend daily concerns of information storage and retention.

Again there is a strong embodied component to this aspect of Angela's phronesis. In particular, Angela believes that the strong relationships with her product team colleagues are preserved and strengthened by the proximity of her workspace to theirs. This proximity enables Angela to, she believes, in effect *absorb* information from her physical and social environment without conscious effort. Consider the following excerpt from our interview, in which Angela describes the physical arrangement of her product team's office space via an extended personal narrative:

> From the day that I got here I was always part of development and I sat with the development team in their project area, and then all of a sudden we started to grow and they were trying to figure out how to seat everybody. . . . So at first they divided us up [i.e., the information developers were given their own area apart from their development teams]. But we noticed that we were out of the daily hub of what was happening with the product. . . . We weren't able to glean as much information just from floating from what you hear or being able to just walk next door and ask a question. And it started to show in the development of the products that we weren't keeping—being kept—abreast of as much as we should have because we weren't there for all that spontaneous discussion and talk and brainstorming that takes place. . . . Right now everybody besides myself has their own place in the information development area. I moved into another area. . . . I'm part of an experiment, which is different from the rest of the people on my core team [i.e., the information development team], and it's proving to be much better. Being able to be with the product team; we know that that's the best way to go.

As Angela notes, the Graphic Forge team occupies the experimental, Petri Dish area of the Information Development Wing (see fig. A.1 in app. A), which consists of several two-person offices surrounding a common team meeting area.

This arrangement enables Angela to find the information she needs quickly by referencing her colleagues rather than searching a computer archive: "So on a continuous basis, if I run across something in the product as I'm using it, I can either IM somebody or get up out of my chair and holler around the door. . . . We have meetings that are right outside everybody's door. . . . So I could just step outside the door." From this, it appears that, like the stories that she possesses, the spaces in which Angela works serve for her as aide-mémoires (Fentress & Wickham, 1992, p. 50). As I noted in chapter 5, ancient memory theory placed particular emphasis on location-

or place-based mnemonics; in a real sense Angela's memory places are actually the interfaces with which and the spaces in which she works: her product's interface, the layout of her office, and the placement of her workspace in relation to her product team. Consequently, Angela does not feel the need to create elaborate archives, to carefully maintain her incoming email messages, to write reminders for the things she needs to accomplish, or to engage in any other mnemonic work. Her stories, her colleagues, and her workspace do this work for her. Or so she believes. Ultimately, however, what seems to occur is that while all or most of the new information, or knowledge, gleaned from these informal "outside the door" conversations may get "entered" (and it is important to note that some of it may not) into Angela's broad-based historical narrative, it does not necessarily get recorded into any formal or "hard data" memory systems such as archived memos or documents entered or added to the memory regimes of any of the communities of practice (e.g., information developers or software developers) or that of the company as a whole.

By contrast, Angela believes that the simple reason the other information developers must rely more on software and other communication tools to maintain team awareness and coordinate group work is principally that they have neither her long relationship with her product nor the physical proximity to their colleagues on those teams. According to Angela, "The other teams are spread out across four buildings. They need a collective way of gathering thoughts. We [i.e. the Graphic Forge team] have lots of whiteboards that have writing on them and, we are kind of the envy of the rest of the company, we've got the huge monitor. We can set it up. We've got movable furniture for moving configurations and tables." In comparison, Peter noted the difficulty in coordinating work with his product teams without interrupting his own work or intruding on his colleagues:

> I spent some time this summer actually over there in the other building. And that was very good when I was working with that team, but at the same time it made working with other teams harder. . . . We use Campfire a lot, which is kind of a team chat room. That happens a lot during the day when anyone has questions—even when they are in the same building and they have offices next to each other—they can't quite talk to each other all the time without getting up. It's just a way to ask questions real quick.

Similarly, Lucy informed me that the only person she tended to routinely consult if she had a question while working was Robert, primarily because his cubicle was adjacent to her own and she could yell to him without leaving her seat. Since they work on the other side of the building, Lucy is far less likely to consult members of her product team when she needs information.

Because studies from a variety of perspectives, including both situated-cognition theory (e.g., Hayne, 2005; Perry, Fruchter, & Rosenberg, 1999) and activity theory (e.g., Swarts, 2004; Spinelli, Perry, & O'Hara, 2005), have explored difficulties related to coordinating distributed teams using software alone, I will not discuss the issue in detail here. Suffice it to say the evidence from these studies points, generally, to the conclusion that physical or spatial proximity adds an informational dimension to everyday interactions and infrastructures that enhances communication and collective and individual memory, and, last, that is hard to duplicate with software alone. In other words, the lack of physical proximity between colleagues adds a need for an additional level of articulation work just to stay organized and on task. In addressing the added challenges that attend working remotely, Brown and Duguid (2002, p.69) point out that "many of the difficulties reflect a misunderstanding of office work, which is too often painted as information handling."

I suspect that anyone who has ever periodically "worked from home" for any length of time can attest to this. One may be "connected" to the office from home via computer, but the atmosphere and even the computer itself both "feel" different; receiving and sending emails may require an additional step; machines like faxes or copiers are in different places (or not present at all); one can't simply "holler around the corner" to ask a question of a coworker. Indeed, even the distractions are different in one's "home office" from the distractions at the regular workplace. As Brown and Duguid (2002, p. 70) further state, "Social systems often play a key part in making even frail technology robust."

In any case, one of the major purposes of Software Unlimited's adoption of the Agile methodology, with its emphasis on daily face-to-face team Scrum meetings, was to maintain just this kind of team awareness and knowledge sharing. However, despite their overall enthusiasm for the brevity and conciseness of the Scrums, the majority of the information developers expressed some reservations about them, indicating that Scrums alone were not adequate to coordinate the teams' work and that the Agile methodology had not been uniformly implemented or adhered to by all the product teams in the organization. Robert, for instance, noted that although he appreciated the team momentum generated by the Scrum, he did not always find it very useful: "It's not the most insightful thing in the world, and a lot of people don't have anything to say really." Similarly, Lucy noted that, because her team's Scrums did not always provide the kind of information about her teammates' progress and activities that she needed, she was sometimes unable to complete her own tasks: "Nobody ever says there are impediments—there are never any impediments [i.e., the third of the "questions" each team member must answer in every Daily Scrum] . . . we do go over stuff but it just comes out like . . . sometimes it's like something should have gotten done, but no one mentioned that they had worked on it or were going to work on it—it seems that there's something missing in this tracking." As with software communication tools, Agile methods do not appear to be as successful at ensuring that information is shared as physical proximity and shared workspaces.

ANGELA USES GESTURES TO "FILL IN THE BLANKS"

As a form of memory practice, gestures may be the most difficult to pin down, scrutinize, and, particularly, to quantify in some meaningful way. Whereas archiving by saving an important document to a specific digital repository or database for later use, or using a reminding practice by handwriting notes about important bits of information in a spiral notebook (or even jotting a schedule reminder on a Post-it), in each case leaves a record of the action or event, gestures by their very nature occur "out there in thin air" and then are gone. From the perspective of the researcher observing and trying to evaluate the meaning, objective, or purpose of gestures as memory practices and also trying, of course, to assess their effectiveness, it is difficult to ask a study participant who has enacted some gesture or another, "Why did you do that?" For one thing, such gestures are often spontaneous, like the hammering motion a carpenter makes with his hand to communicate his immediate need to his apprentice because for the moment the word "hammer" happens to have escaped his tongue. For another, such gestures tend to be abundantly clear in their simplicity, as in the same example of the carpenter, and as they must be in order to be effective. One need only think of the game charades to grasp the idea that the simpler and the more pinpointedly targeted the gesture is that is performed by one player, the more effective the communication of the puzzle will be to the other players, leading to winning the game.

The difficulty in quantifying or "capturing" gestures as a way in which we deal with information as memory practices, however, in no way diminishes their importance for understanding the thought processes behind them. McNeill (1992, p. 12) puts the vital nature of gestures succinctly:

"These gestures are the person's memory and thoughts rendered visible. Gestures are like thoughts themselves. They belong, not to the outside world, but to the inside one of memory, thought, and mental images." However, when I first observed Angela's gestures, specifically those which occurred in the second observation session in connection with her meeting with Carl, I found them to be peculiar, but certainly emphatic enough to make note of in my research journal. As I described them earlier, in disagreeing rather strongly with Carl's position, Angela began to make pointing or jabbing motions toward Carl's computer screen with her pen, as if to, as I said, puncture some sort of invisible barrier between her viewpoint and that of Carl, or perhaps in an effort to poke holes in Carl's argument. And at first, I wasn't sure what to make of this behavior. However, upon reflection, I can now offer some valuable insight into what was going on with Angela in this exchange.

To understand the dynamic underlying this scenario, we must first look at what Angela brings to this meeting, which ultimately turns into a mild confrontation of sorts. Recall from chapter 4 that, for the meeting with the software developer Andie, Robert brought a sophisticated and laboriously prepared reminder document containing screenshots and a wealth of solid reasons backing up his position; in other words, Robert brought to that meeting all of the warrants he could muster as a means of ensuring that he would win the debate. By contrast, Angela brings none of this to her meeting with Carl—in fact, she goes into the meeting armed with nothing at all beyond her story, the historical narrative, and perhaps her own perception of her stature and authority by virtue of her tenure. She brings no documentation, no paperwork: more particularly, she brings no company-based, authoritative instantiation or pronouncement of the "corporate" memory regime, such as a company policy statement, to back up her position in the argument. Her attempted rebuttal to Carl instead consists merely of insisting that she has "talked to users" and that these users "like" the new graphic-based help, and that she knows the "community at large" is on her side. This is a weak argument in the absence of any sort of either supporting documents or policy issued by the company or, at minimum, some sort of reliable user evaluation data, perhaps in the form of accumulated user reaction to the new graphic-based help screens that Angela favors.

Here is where Angela's gesturing practices come in. By pointing and jabbing at the screen, Angela is not trying to metaphorically "poke holes" in Carl's argument: What she is actually doing is trying to demonstrate or emulate what the program users are doing, how those users are using or navigating the program on the screen, or the help graphics, or whatever else they are viewing onscreen. Not unlike employing manual tools, working with the aid of a computer screen involves multiple cognitive and physical systems—the mind thinks and decides actions based on what the eye sees on screen, and the hand moves the mouse to move the cursor to perform those actions, just as surely as the carpenter's mind sees the two-by-four that needs to be nailed and directs the hand to wield the hammer to hit the nail to perform the action. As McNeill (1992, p. 13) states, "The gesture reveals not only the speaker's memory image but also the particular point of view that he [has] taken toward it." When the carpenter motions to his apprentice, his "point of view" is, "I need the thing that I use like this," swinging an arm with a clenched fist on the end of it imitating the holding of the tool. When, quite similarly, Angela is pointing at the screen with her pen, her point of view is, "Users need to use the [mouse, i.e., move the cursor] like this." In the absence of archival, hard-copy documents or, for example, printed exhibits, such as screenshots with, say, arrows or other devices to show how users prefer to operate the program by navigating the screen displays, Angela is using gestures as a memory practice to "fill in" the physical, mnemonic component of what users actually do when they operate the program or access the graphic-based user help screens—and especially to try to communicate all of that information and knowledge to Carl. Thus, what we started out with at the beginning of this section, that is to say, characterizing gestural memory practices as being somewhat vague, spontaneous, and difficult to quantify or evaluate vis-à-vis their effectiveness, it turns out that gestures, at least in this instance, display with particularly clarity their capacity to enable technical communicators to perform their job function as described in the very beginning of this book: taking information, turning it into knowledge, and communicating that information to others, both in the community (or communities) of practice and in the organization at large.

Finally, if Angela's speech in the context of this episode with Carl is, in and of itself, weak or insufficient, and ultimately unconvincing (e.g., "I know what users want"), and likewise her gestures, taken by themselves are, let's just say, difficult to decipher taken on their own, McNeill (1992, pp. 13–14) aptly explains why this is so: "Jointly, speech and gesture give a more complete insight into the speaker's thinking. . . . If we were to look only at the gesture or the speech, we would have an incomplete picture of the speakers memory and mental representation. . . . It is only through a joint consideration of both gesture and speech that we see all the elements."

In closing this section however, I might make one other observation about Angela's approach in her meeting with Carl that leads to an intriguing interpretation of her motivation. Specifically, Angela's concerns reflect both her storied flair for creative, innovative, and pleasing design and her feelings of proprietary ownership over user advocacy to ensure user satisfaction.

Summarizing Angela's memory work, both Angela and the other members of the information development team appear to believe that Angela has achieved something very near to a state of phronesis, the perfect mastery of the knowledge of her workplace. This mind-body union with her work environment—comparable in some sense to the mastery of his piano achieved by Derek Paravicini—enables Angela, in this narrative, to seamlessly recall information and effortlessly invent knowledge in response to the kairos of any situation that could conceivably arise in her workplace. This not only reduces the amount of time Angela must dedicate to performing memory work; it also, she believes, makes her an innovator, a creative force on the information development team: "There are many people on my team that are . . . interested in . . . content management and structure. I'm more on the 'let's be creative' side." Last, Angela's apparent phronesis contributes significantly to her ethos and authority with the other information developers, who repeatedly indicated that they held her abilities in awe and aspired to achieve similar levels of knowledge. All of this would be wonderful for Angela and her teams, if only it were true.

THE POLITICS OF MEMORY

And yet . . . For all the praise the information developers heap upon her and for all of Angela's own apparent confidence in her ability to "just from experience kn[ow]" the things she needs to know, analysis of the observation sessions suggest that a different interpretation is also possible. When Angela's memory practices—and in particular, her predominant reliance on the memory practice of storytelling—are considered in comparison to other accounts of the other actors, a more nuanced picture of the effectiveness of Angela's practices emerges. In short, Angela's state of mastery may not make her as effective at meeting the needs of kairos or securing her ethos as she and the other information developers believe, or as they may want to believe. Taken together, these accounts suggest that mastering memory, the condition of phronesis, is not a fixed point, a final stage of achievement, but, rather, always an ongoing process in a memory regime that is itself continually evolving. In other words, Angela's possession of the stories and her sharing of these stories with other members of communities of practice can create problems as well as solve them.

STORIES AND THE CHANGING CONDITIONS
OF PHRONESIS

To understand this, let's take a closer look at what is perhaps the most striking aspect of the training meeting, which arises from Angela's obvious agitation at Carl's mild reproving of the online help she has created for the upcoming release of Graphic Forge. At a number of points later in the study, I attempted to learn from Angela what it was about this meeting that had so agitated her, but in each instance I found her reluctant to talk about the incident. However, despite her reluctance to talk about the incident directly, she made several offhand comments over the next few weeks that indicated that the incident still rankled, as when she quipped, "If we were training [i.e., the Training Department], we'd have to mind map each card before we put it on the board first," during a usability activity she participated in with the other information developers. More tellingly, despite her initial enthusiasm for my study, after the incident with Carl, Angela became far more reticent and no longer encouraged me to attend her meetings with other members of her product team.

At one level, of course, the cause of Angela's change in attitude seems obvious—she was probably embarrassed that an outside observer had witnessed her work being critiqued. At another level, in light of Angela's stature with the company, her depth of knowledge about the company's products, and confidence in her own memory practices, her behavior appeared somewhat puzzling: why did she care so much if a new employee who was ignorant of the thinking that had gone into her designs should attempt to offer his advice? She clearly knew more than he did and handily won the argument by invoking her superior knowledge and understanding of the company and its users—"why it is the way it is," as Lucy put it. So, why did Carl's criticism appear to sting? What was the source of Angela's apparently unwarranted defensiveness?

Vickers and Fox (2010, p. 902) point out that storytelling is "a situated practice, the reasons for telling stories always local and occasioned in situ." Storytelling is also, as I note earlier, a rhetorical memory practice through which the past is used and interpreted for present circumstances. But the story itself is not static in the present moment. The problem with the historical narrative of storytelling as a memory practice is that the answer to the question of why things "are the way they are" today, or at the moment, is not always a good—by which I mean an appropriate or suitable—answer to the question of whether those things should stay that way. In short, the bases of phronetic knowledge—and the condition or state of phronesis to which this knowledge leads—do not appear to be fixed and static, making phronesis a "final" stage of achievement, but rather are always part of an ongoing process in a memory regime that is itself continually evolving. This is particularly true in a fast-moving, innovative technology company like Software Unlimited. When we analyze more closely the interaction between Angela and Carl, we realize that Angela did not really "win" the argument by force of logic or based on the superiority of her tenure-seasoned knowledge (or even on the superiority of her redesign of the online help pages for Graphic Forge 5): She won it largely through intimidation based on that longer tenure and history, and her perceived position of knowledge and authority within the company.

Also, in order to be effective as an integral part of a memory regime, a story needs to be constantly retold, perhaps in part because of the fact that both the story and the regime are continuously evolving—the story must therefore be updated and retold often if it is to be effective in molding the memory practices of others in the organization. In the training meeting, there is really no time for Angela to recite chapter and verse. She effectively, if abruptly, ends the discussion, but it is doubtful that she has won Carl (or any of Carl's team) over to her way of thinking.

A misunderstanding that occurred a few weeks after the training meeting incident offers yet a further indication of where the source of Angela's discomfiture at Carl's critique may lie. A representative from the Marketing Department had asked Angela to create a document promoting the new and innovative features that she was creating for the online help system in the upcoming release of Graphic Forge. The marketing team wanted a document that it could take to upcoming trade shows that would introduce customers to the new features of Angela's graphic-based online help. With her penchant for creative efforts, Angela had created a colorful and "fun" poster-sized document that she believed would provide an interesting and engaging introduction to both the product and the online help. In addition to being displayed at trade shows and included with beta versions of the product, Angela believed the poster could also be boxed with the final product when it shipped. After completing the draft, Angela hung the poster on a bulletin board in a common area of the building and sent an email to all employees requesting that they stop by to view it and give her feedback on it.

At the regular information development team meeting the following week, however, Angela informed the information developers that, rather than providing the sort of constructive feedback she had hoped for, the marketing team had made derogatory comments about the poster and had even accused her of deliberately ignoring the new brand identity they had created for Graphic Forge. Because of the poster, Marketing now thought the information developers were "doing something different" from what the rest of the company intended. Angela wanted to create a document that was "fun and arty," and Marketing had expected a document that "was plain, white, and simple," so that it would be comparable to the marketing materials of Software Unlimited's chief competitor—and presumably, *competitive* with the rival company's product line. Consequently, the Marketing Department cut the information development team out of the loop, over the intervening weekend, by putting together and printing its own document to take to trade shows and pack in with beta versions of Graphic Forge. Unfortunately, from Angela's perspective these marketing documents, which would now be shipping with beta versions of the product, were "not engaging or interesting."

What may be at work here, at least in part, is the sort of conflict that I discussed in chapter 3, among two or more of the subcultures within Software Unlimited, that is, among two or more of the communities of practice within the overall organization, and quite possibly the beginnings of the inadvertent creation of "silos of information." For example, the largely newly formed Marketing Department is focused, among other concerns, on creating a new or revitalized brand identity for the new version of Graphic Forge, and also on specifically targeting the promotional materials of Software Unlimited to go head-to-head against its competitors—both issues that probably did not enter into Angela's thinking or her conception of the task, either now or especially in the past. While Angela's creative emphasis might be on "colorful, fun, and engaging," as well as on user appeal, Marketing's emphasis, quite understandably, appears to be on competing successfully in the marketplace to maximize sales. In many businesses, this is a classic confrontation. And we can learn more from this example by examining the tension it created both within and across the subculture communities of practice within the company.

The information development team manager, Becky, understandably, appeared to sympathize with Angela's position, and, moreover, to consider the incident an affront to her team's expertise: "Everyone thinks they can write." Becky explained to the rest of the team that the misunderstanding had occurred because marketing requested the document on very short notice and that the problem had then escalated because upper management was out of town and out of contact over the intervening weekend, both fairly weak excuses for a company that is built on speed of innovation. Nevertheless and stalwartly, she contended that if they'd had another twenty-four hours, she believed they could have resolved the issue, but, instead, "we now have thirty thousand crappy posters, and then we'll be back on track."

Last, attempting to sum up the lessons learned from the experience, Becky noted that the information development team had "learned a lot about the relationship with Marketing," and that from now on they should "overcommunicate" when dealing with the Marketing Department. Significantly, as part of this over communicating, Becky advised that the information developers "give the backstory," because "with 170 employees versus 20, there are larger chains" and "people know less than you think." It is clear that this advice is mainly targeted at Angela: Angela is the information developer at the center of this dispute, and, even more tellingly, Angela is the only member of the team to have experienced the firm when it consisted of just twenty employees. In other words, although she is addressing the entire team, Becky is attempting to convey to Angela in particular that she can no longer assume that other employees already know her history and reputation. Instead, Becky' asserts, Angela must retell her story (which contains the backstory of the Graphic Forge product) if she expects to continue achieving her objectives in this newly enlarged company. More important, however, is that Becky's overture is a clear indication that Becky, at least, realizes that the expansion of Software Unlimited is changing the dynamic of authority within the company. At the very least, it is obvious that Becky realizes that as the company adds personnel, fewer and fewer of the employees (as a percentage) respect, recognize, or might even accept Angela's authority (particularly those in other departments), and, as a consequence, Angela's stature is dropping, as is the validity and effectiveness of her story. Hence the need to constantly reinforce it by retelling it!

A principal source of Angela's vexation at the new employee Carl in the training meeting, then, turns out to be not so much Carl himself as what he represents: a rapid and exponential growth that threatens to leave the old guard behind. In fact, in addition to the incident with Carl, Angela references the company's growth no fewer than six times in our initial interview, noting that lines of communication between teams becomes more difficult: "As we grow, things get more complicated"; that projects grow more unwieldy to manage: "So it always gets complicated as we grow, not so much as the product moves on"; and even that finding a place to meet now requires approval "There's lots of all of these policies get put into place when you are growing up. Before it was, just find a place to meet. And now we've got designated areas." The subtext to Angela's dissatisfaction and anxiety over this emerging state of affairs is that it is far easier to retell, repropagate, and reinforce the bedrock story of "why things are the way that are" when your organization consists of 20 people in a couple of rooms or a small office suite than it is when the company has 170 employees spread across several offices in multiple detached buildings, at which point, in fact, it actually may be impossible to do so as a basic matter of practicality. At some point in this rapid development and growth, presumably, Angela's story ceases to be embodied in the overall corporate culture and the memory regime of the organization, and her retellings of this story have, as a means of managing information, ceased to contribute to the propagation of phronetic knowledge in either herself or her colleagues. For Angela, storytelling appears, in other words, to have ceased to function as an effective rhetorical memory practice.

I would suggest, therefore, that Angela's problem with the recent changes that have taken place at Software Unlimited is not that the company has grown so large that she no longer recognizes *it* but that it has grown so fast that it no longer recognizes *her*. As a consequence of this, Angela increasingly finds herself in situations where she must rearticulate her ethos by either referencing or retelling the story she told me in our interview (i.e., the first observation session). When newcomers or novice members of one of her communities of practice like Carl challenge this authority, Angela is able to merely reference her story (e.g., "I've talked to users"), but, when challenged by a team as influential as the new Marketing Department, the only thing that will suffice is a full retelling of the "backstory"—and even that may not completely do the job because of other legitimate concerns that represent specific challenges that (as in this example) can only be met by the Marketing Department itself, such as battling the competition or redefining the company brand. At some point, by clinging to a historical narrative that may be petering out, Angela risks being seen by the other individuals as in effect saying, "Things are the way they are because I say that's the way things are."

STORYTELLING AND THE CHALLENGES OF KAIROS

Knowing when, where, and with whom (i.e., questions of kairos) to share one's story can also be difficult. This is illustrated by considering another controversy that Angela became embroiled in during the study, this time with one of her own communities of practice. In the months after the poster controversy, it became increasingly clear to Angela that there were several major problems with the user interface for the upcoming release of Graphic Forge: the new version radically changed the terminology employed in previous versions, the icons on the primary user interface were too large, which required users to deploy multiple confusing pop-up menus in order to navigate the functions in the application, and the workflows were too complex, requiring users to perform multiple steps to accomplish tasks that had been simple in earlier versions.

Acting in her role as user advocate, Angela had attempted to intervene at several points in the development process, but, despite her extensive experience with the product and with the product team, she had made little headway. I observed an instance of this during the product team's Daily Scrum on September 18, when Angela (possibly violating the rule of Scrum that discourages raising issues unrelated to immediate needs) voiced her concern that the UI icons were too large and therefore "inelegant" compared to the icons in the Windows operating system on which the application runs. Ben, the representative of the UX department serving on the Graphic Forge team, was not receptive to this criticism and informed Angela that the larger size had been chosen deliberately in order to "give users as much information as possible" on the main interface, adding that the standard 16 × 16 pixel Windows icon did not provide enough detail. Angela listened to this explanation and appeared to acquiesce without further comment.

Behind the scenes, however, Angela began to recruit allies to help her make her case. Most notably, Angela enlisted Becky. To build support for their case, Becky asked the other information developers to participate in an informal usability evaluation of the new interface at their December 5 team meeting. Angela was notably absent as the meeting began. Becky informed the information developers that they themselves represented the closest thing to "power users" of Graphic Forge at the company, so their feedback would be especially valuable.

Lucy began to test the application while Robert, Monica, and Peter watched her do so on an overhead display screen. As Lucy worked, she and the others noted that the menu system did not appear to support user tasks very well (e.g., "everything is spread out"), that simple tasks required "too many steps," and that one feature was "really awful." In response to the preponderance of negative reactions, Becky replied "you are warming my heart." When Lucy and the others finally called a halt to their testing, Becky informed them that she had asked Angela not to attend so that she could get their unbiased opinion. She noted that she and Angela had been going "round and round about these problems" with the product team manager to little avail, so she hoped this feedback from expert-level users of previous versions of the product would serve as "ammo to make a case against the new UI and workflows."

It is a matter of conjecture whether the reason that Becky gave for excluding Angela from this private usability evaluation "beta test" is true or not. One may speculate that the real reason was to avoid any intimidation of the less experienced and shorter-tenured information developers on being confronted with the very real prospect of having to disagree with Angela—face to face—who is, after all, the most senior "old-timer" in their community of practice and, in some sense, the team leader emeritus, now that Becky is manager and head of the department. Alternatively, one could speculate that excluding Angela from the test was Becky's way to avoid hurting Angela's feelings or embarrassing her in the event that such a disagreement might actually arise; after all, Angela has been a respected and much-liked mentor and friend to all of them, once again including Becky, even though she technically is now Angela's boss. Speaking objectively, and based on all of the indications in my research, I would have to surmise that all of these factors played at least a small part in Becky's decision to exclude Angela from the usability evaluation test.

It is important to note that all of this activity, as well meaning as the intent might be for the good of the product, serves to pit departments within the company against each other, and thus works against mutual cooperation toward the achievement of the company's goals. In stark terms, naturally none of this "behind-the-scenes" activity was reported or talked about beyond the information developers—at least not openly that I observed—and as such, it stayed within their community of practice, for reasons that are all too obvious. Further, the beta test that Becky set up intentionally excluded even one of her own team members—and the most authoritative, senior member at that. (As stated earlier, it is possible that Becky's intention was, in fact, to elicit the unbiased opinions of each of the other members of the information development team with Angela not present. What seems more likely, however, is that Angela exerts a certain level of intimidation over even her own team members, which Becky recognizes, and which would make the less senior members reticent or even outright unwilling to disagree with Angela. We have already seen how negatively Angela reacts when her view of things, her story, is questioned or disputed. Ultimately, this may be yet another indicator of why Becky is the department manager.)

Ultimately, however, all of these extracurricular machinations serve to further support the assessment expressed in chapter 3 that multiple subcultural memory regimes exist within the structure of Software Unlimited, and that these regimes are organized along new and evolving departmental lines as different and separate communities of practice. As I noted in chapter 3 and elsewhere, while this kind of loose, satellite corporate structure may encourage innovation and creativity by individuals within or in con-

cert with their particular communities of practice, it may also lead to conflict as a result of organizational fragmentation.

In any case, despite Becky's beta-testing session, as well as some other
user advocacy efforts over the next few weeks, Becky and Angela's attempts
to get the interface and workflows changed were largely unsuccessful. In
fact, their efforts were so unsuccessful that, during a January team meeting,
Becky asked the team to begin brainstorming possibilities for a new team
name and statement of identity that would better convey to other teams
the role that the information development team is supposed to play as user
advocates. Her failure to achieve traction with the Graphic Forge team had
convinced Becky that other teams needed to keep in mind that "user assistance is more than just documentation."

At this point and in this context, Angela's story makes its last appearance in these pages. When Becky finished explaining the focus of the meeting, Angela recounted for the information developers an impromptu meeting she had called with the Graphic Forge team earlier in the week in which
she was finally able to narrate her story in full for all the members of the
team, both new and old, to hear (the regular Daily Scrum meeting having
proved to be, by design, a poor place to attempt to tell an extended story).

Expanding on the version of the story she told in our interview, in the
first observation session, and hearkening back to her earliest days at the
company shortly after she was hired by the CEO himself, Angela explained
to her product team that "watching and being keepers of the work process
flows is part of our [i.e., the information development team's] job." This
assertion elicited an interruption from Ben, the UX representative on the
team. Ben countered Angela by arguing that creating and maintaining the
user interface was his job, not the information developer's. In reply, Angela
explained to Ben and to the team as a whole that her original title at Software Unlimited had not been information developer, but, rather, process
improvement manager and that she was brought on board "not to write
help files but to move from shareware model to commercial model." She
further explained that, because her process improvement work familiarized
her with the workflows, she was in a good position to write the online help
and other documentation. This, then, was the origin of the information developer role at Software Unlimited. The team was created and continued
to exist primarily to improve user work processes and only secondarily to
write the documentation, a welcome evolution of the traditional responsibilities of the role of the technical communicator, but an evolution that, as
can be seen from the information development team's conflicts with other
teams, has not been fully supported by management. Angela concluded her

retelling of her story by informing her product team that "a help file is like a report card for the application—if the work processes are good, the help file is short and simple and easy to explain."

In this account, Angela is finally compelled to create special conditions in which she can retell her story in all its glory—a non–Daily Scrum team meeting—in which she is able to mobilize the intertwined history of professional growth and product evolution—the "three-year reason" as Lucy might put it—in an attempt to achieve the same type of user advocacy goal that, as we have already seen, her junior colleagues have been struggling toward.

Angela's is a compelling story, and one not without its lessons to be learned. Graphic Forge has evolved to meet her needs as the first "prototypical" user: "One of the goals that has been in place since I started with Software Unlimited was design a Graphic Forge product that 'Angela' can actually use." Thus, it is significant that retelling her story in full ultimately proved incapable of swaying her product team colleagues or their manager, and consequently Graphic Forge ended up shipping a few months later with a UI and workflows essentially unaltered from the problematic ones flagged by Angela after the September 18 Daily Scrum meeting, and more specifically identified at the information development meeting the previous December. Like the incident with the poster, the tight deadlines created by the ceaseless forward momentum of the Sprint cycle, which also made Robert and Lucy's working relationships with their product teams difficult, fostered a memory regime that was increasingly militating against the sorts of storytelling practices that an earlier, smaller incarnation of Software Unlimited had tolerated and encouraged as a means of knowledge sharing. Lucy had earlier speculated that this might be the case: "There was a lot of stuff like that around here—you know just being a small company with all this stuff it was like all these reasonings. . . . I don't know if we are kind of getting away from that as we get a bit bigger." It took the release of the company's oldest product to prove this speculation true. In sum, because Angela's storytelling proved inadequate to the kairos of the situation, the product, like the poster, shipped without the benefit of her knowledge and expertise with respect to the needs and preferences of the company's users.[4]

CONCLUSION

Angela's case offers important insights into the role that effective rhetorical memory practices play in achieving phronesis and into the research questions that frame this study. Specifically, Angela's case suggests answers to

the "Why" questions central to phronesis: why does Angela and her colleagues' memory work matter; and why, if at all, does it enhance their
status with their colleagues?

First, there is the issue of hexis, or "'complete mastery' of subject and
self" undergirding the condition of phronesis (Carruthers, 1990, p. 178). In
several of her colleagues' accounts and in her own, Angela is portrayed as
a paragon of this kind of expertise, an expertise so great that it seemingly
transcends the everyday sort of mastery that comes from advanced technical skill or from achieving seniority in a community of practice. That is,
Angela is often portrayed not only as a full participant in her community,
but also as a sort of ur-participant or ur-user around whose needs and desires
her product, her product team, and even her and her team's material and
physical workspaces have been built. This state of total mastery ought to
enable her to find and use information to achieve her job goals effortlessly
with little need for the kind of careful memory work that the other information developers require.

Yet the actions and interactions in the scenes and incidents recounted
above reveal that such a state as Angela describes is illusory. That is, unlike
achieving full-membership in a community of practice, achieving phronesis
is not a plateau or fixed status at which one can, through extensive training or long association, eventually *arrive*, but rather a "flexible and open
ended" disposition or condition that requires ongoing and continual practice, *rhetorical memory practice* to be specific (Dalton, 2004, p. 613). Angela
possesses the backstories of her product, but to succeed she must continually retell or rearticulate this story for new audiences in new situations. She
cannot rest on her laurels by letting her reputation precede her; in "firms
pursuing an innovation strategy" like Software Unlimited, there are always
new products to master and new audiences, both internal and external, to
persuade (DeLong, 2004, p. 31).

At the same time, Angela's failure to utilize some of the other memory
practices identified in the course of this study significantly works against
her, as well as against the "currency" of her story. In particular, by not finding and/or archiving new information (e.g., by ignoring and not organizing
her emails, for starters), Angela very likely fails to keep abreast of new technologies that might need to be incorporated into Graphic Forge (or any of
the other products offered by Software Unlimited), she may fail to keep in
touch with changes in user preferences or trends as a result of that emerging technology (or as a result of simple demographics), and she most certainly misses out when it comes to keeping up with what the competition
is doing. Indeed, her extreme confidence alone, in her story would seem to

incline her to be oblivious to what the competition is doing and to ignore it, as arrogant as this may seem.

Even more disturbingly, by ignoring her emails Angela almost certainly fails to keep herself informed about developments within her own company. Given the fact that the company's personnel are now spread out across multiple offices and locations, it stands to reason that the vast majority of interoffice and interdepartmental communications are facilitated via internal company email. Realizing this, one has to wonder, again with good reason, whether some of the conflicts recounted in this chapter could have been easily avoided if Angela had simply read her emails! (And further, she might have even had a better chance of getting her way if she had headed off the criticisms, in the appropriate forum, by routinely answering relevant emails as a matter of course, rather than having to do so at inappropriate times [under looming deadlines], in inappropriate circumstances [e.g., a Scrum meeting meant to be short and sweet—no time for storytelling], and finally, poorly armed with only her story [e.g., no archive-based, authoritative, company-endorsed documentation or policy statements].)

I would suggest, therefore, that Angela's reaction to Carl's critique of her redesigned online help system, her unwillingness to more forcefully articulate problems she perceived in the product UI at the September 18 meeting, and her failure to assert her expertise to her teammates until mid-January of the following year may be as much a result of her own complacence as they were of an aggressive shipping deadline or relentless forward momentum of the Agile methodology. The source of this complacence was that Angela and the other information developers maintained the false assumption that her reputation alone could persuade her teammates and steer the product's development: she knew everything there was to know about the product UI, the product had been built with her in mind, she was a dynamic and innovative member of the team, so what could possibly go wrong? Well, as might be surmised from this analysis, the danger in all of this—the reliance on storytelling coupled with the failure to access and incorporate new knowledge and innovation into one's hexis—is the very real probability that Angela's story will eventually morph from the company's story into what might be referred to as "company lore"; in essence, the way "things used to be done."

In other words, over the preceding years Angela's story had become part of the memory regime itself. In fact, in the early days when she first joined the firm, and it consisted of some twenty people, Angela's historical narrative might have equated or corresponded directly to the overall, company-wide organizational memory regime itself—a time when that regime speci-

fied that the canonical practice for finding out "how we do things" or the
sound reasons why things "are the way they are" was quite specific and
clear: "Ask Angela, ask Angela, ask Angela." Unfortunately for Angela and
for the information development team as a whole, which continues to some
degree to view Angela's story as a canonical part of its satellite regime, the
problem is that the corporate regime has continued to evolve and change,
while these assumptions have not. As Deuten and Rip (2000, p. 80) note
in the quotation that opens this chapter, by "telling a story in which you
are a character yourself . . . you cannot permit yourself too much deviation
from the expectations connected with this role." In Angela's story, she was
both the first user of the product and the creator of the initial workflows, so
why would she not be listened to? Why should she have to repeat her story
again and again? In short, "narratives create inertia," and, as a consequence,
in Angela's case, concerns of hexis eclipsed concerns of kairos (Deuten &
Rip, 2000, p. 77). That is, rather than continuing to grow and learn or to
cultivate multiple hexeis—the path to true creativity, according to Hawhee
(2004)—Angela had become habituated to doing things the way they had
always been done, and, in doing so, had strayed from the phronetic state
that had earned her her earlier reputation.

As a second point, this discussion of phronesis and hexis invites us to
revisit the relationship among communities-of-practice research and rhetorical theory. As Fox (2000) notes, communities-of-practice theory views
learning and mastery as distributed. Communities-of-practice theory
posits that learning and mastery occur within the community: in a well-
functioning community, when one individual member acquires new knowledge, he or she shares it with the other members of the community, thereby
enabling the community to achieve more than it was able to previously. On
the other hand, the rhetorical concept of hexis, as an "assured facility . . .
that transcends the rules themselves," and with its emphasis on embodiment, appears at first to be concerned only with individual achievement
and, therefore, to be incompatible with the view that learning and mastery
are distributed (Carruthers, 1990, p. 69).

However, Angela's case demonstrates that this is not so; at least it is
not so in the context of a community of practice, where knowledge must be
shared in order to count as knowledge. The source of the hexis that Angela
and her information developer colleagues believe she possesses lies primarily in her stories, but, like any other archived information, these stories can
only become knowledge when they are used: Angela must tell her stories
to her teammates. Moreover, the kairos must be right. As any professional
storyteller knows, to achieve its intended goal, a story must be told at the

right time and in the appropriate place. However, the growth of the company has significantly militated against both of those goals: to tell a story, and to tell it at the right time and place. Both the burgeoning population of employees and the outspreading geographical locations of those employees has seriously curtailed the opportunities for Angela to tell her story. Although Angela perhaps should not bear full responsibility for the flawed product that eventually shipped, her failure to advocate for users until too late in the design cycle contributed to the collective failure of the community of practice to achieve its aims. More fundamentally, however, Angela's failure may have much more to do with utilizing other memory practices that would have enabled her to better meet the requirements of kairos.

Last, Angela's case—Angela's story—suggests the answer to the "Why" questions of the study: Why does memory work matter? Why does it enhance the status and prestige of technical communicators in their workplaces? The answer is that memory practices matter because, when they are utilized effectively, they translate and share information; they make knowledge for both individuals and communities. And memory practices are particularly important for technical communicators because, as each of the information developers' stories illustrate, in high-tech firms like Software Unlimited, these practices are often the only means by which writers and communicators can acquire phronetic knowledge sufficient to warrant and win arguments with other, often more powerful, communities within their organizations. In other words, memory practices do political and rhetorical work, and effective memory practices do this work well.

CHAPTER SEVEN

The "New" Art of Memory

Technical writers, then, must create order out of a chaos of data. . . .
Writers don't just sort data; they must educate their readers—the different layers of an organization, not just the end user—by using creative
methods of presenting new concepts, products, and technologies.
—Andrea McKenzie, 2008, p. 20

I will now return to the research questions that motivated this study and
attempt to synthesize the answers that this study of Software Unlimited
has suggested:

What types of organizational information are the most important for
technical communicators?
Where does this information reside and how does it move through the
organization?
How do technical communicators transform information into useful
knowledge?
How does this memory work contribute to the status and professional
identity of technical communicators within their organizations?
Why does memory work matter; that is, why does it enhance the status
and prestige of technical communicators in their workplaces?

In particular, I want to focus on how the concepts and framework offered
by this book—concepts from rhetorical theory, communities-of-practice
theory, and the concept of the memory regime (more particularly, the
more sophisticated perspective of the multiple memory regimes of innovative companies or organizations)—answer these questions in ways that
enable us to envision what the new art of memory looks like. I also want

to examine how "new" that art of memory is, and it is important to keep in mind that I have grounded these social theories upon the foundation of rhetorical practice, drawing principally from Aristotle's broader concepts of knowledge, practice, and expertise.

MEMORY WORK IS A NECESSARY BUT INSUFFICIENT PREDICTOR OF SUCCESS

The three information developers on whose memory work chapters 4–6 focus are representative of common stages of membership in communities of practice.[1] Robert is a newcomer, Lucy is experienced but marginalized, making her something rather like a midstatus journeyman on her product team, and Angela is an old-timer, the team member who believes that she has seen it all and who knows in which closets the skeletons are kept. The three also display what appear at first to be highly idiosyncratic sets of approaches to memory work, but which, upon further consideration, turn out to be, in their broad concerns at least, actually fairly typical and expected responses to their individual situations. In retrospect, it strikes us as not too surprising that a newcomer as savvy and ambitious as Robert would immediately set about accumulating as much information about his new workplace and its products as he can; or that an employee who is struggling, like Lucy, to achieve an appropriate work-family balance would be hyperconscious of the need to the squeeze maximum productivity out of every working hour and, equally importantly, of the need to have this efficiency recognized by her peers; or, finally, that an old-timer would rely more on inductive reasoning based on past experiences—and a self-perceived ability to assimilate new informational developments into the stream of her reasoning—than her newer colleagues, as is the case with Angela.

Each of the information developers regards these practices as, in various ways, helping him or her respond to the kairos of daily work situations: carefully maintained archives help Robert and Lucy author documentation with ease and provide a stockpile of information that they can use as evidence to support their user advocacy work; scrupulous reminding practices ensure that Lucy stays focused and ready to make valued contributions during team meetings; and an extensive stockpile of the backstories and product history enables Angela to perform her duties without excessive preparation, metaphorically reminiscent, to a degree, of the place/image mnemonic of the ancient orator-rhetorician that I alluded to in chapter 2. Each of the information developers also views these individual practices as accretive, as contributing toward *something*, something larger than simply meeting individual needs or immediate exigencies. This something is, I have argued, one of two states of expertise defined and described in this book in Aristotelian terms. The first plateau of expertise is techne, a status or state of preparedness in which an individual is ready and has on hand the resources to meet any exigency that may be encountered during the routine execution of his or her job duties. Many if not most of our memory practices are directed toward this type of kairotic readiness, and, in the context of this study, Robert's practice of keeping a "supporting materials" archive containing things like "just three sentences or two different sentences that have pros and cons, so that in one breath of air you can explain it" serves as a typical example.

The second plateau is phronesis, which, as I described in chapter 2, encompasses a degree of mastery that entails not simply knowledge of the technical details of a practice, but also a deeper, fundamental understanding of the social and ethical reasoning that underlies practice. Phronetic knowledge, I asserted, is knowledge that includes a social awareness component, and that contributes, consequently, to the acquisition of a good professional identity within one's organization and communities of practice. Necessary to this level of knowledge is the embodied understanding of and commitment to one's métier described by the concept of hexis: phronesis is always embodied. When they are most corporeal and oriented toward the long term, our memory practices are directed toward phronesis. With their evocative, corporeal titles, Robert's "back of the mind" archive and Lucy's "data dump" (as well as her copious handwritten notes detailing the "reasonings") are good examples of memory practices directed toward achieving the sort of deeper, embodied understanding entailed in phronesis.

In her belief that she has achieved such a memory-ready state, Angela serves as both an example and a cautionary tale. By virtue of her lengthy tenure with the firm combined with her embodied (physical, mental, and even emotional) proximity to her product, her colleagues, and her tools, Angela appears to have achieved a state of phronesis and can now operate independently of the kinds of rigorous memory aides on which others rely. Angela, for her part, appears to have bought into this narrative (e.g., "I'm able to just know") and, as a consequence, has come to believe that she is exempt from most forms of memory work (e.g., "Other people, they have file folders and they keep track of things. I just don't").

Angela can make this assertion because the memory culture—the memory regime—of Software Unlimited has traditionally encouraged just such an informal, unsystematic, oral approach to preserving its most important organizational memories. As Lucy puts it, knowing the backstory "makes a difference around here." Nevertheless, like many other firms before it,

Software Unlimited is finding that as it grows it needs more formal and systematized methods for preserving, accessing, and sharing information. The ongoing implementation of Agile and the Daily Scrums across the organization represent a major attempt to do a better job at this memory work.

For her part, Angela is finding that these changes are making her traditional practices less effective. In other words, she is finding that, for knowledge workers such as her, at least, there is no such thing as a true state of ultimate preparedness. Wenger (1998, p. 101) points out that generational discontinuities do arise within communities of practice: "Communities of practice are not havens of peace. . . . Their evolution involves politics of both participation and reification. Generational differences add an edge to these politics by including the distinct perspectives that successive generations bring to bear on the history of a practice." Achieving phronesis in the workplace is, to be sure, a desirable goal for an ambitious person wanting to make a significant contribution to his or her organization. Moreover, on strictly the level of personal satisfaction and of having a sense of well-being or self-worth, we would all like to see our working lives operating like the proverbial well-oiled machine, wherein our contribution is valued and appreciated. But phronesis is, nevertheless, not a condition that has a stopping point, at least not for information-managing knowledge workers. While it is certainly not the purpose of this study to judge the job performances of the information developers who graciously agreed to participate in it, it is useful and illustrative to draw some general inferences about the levels of expertise shown by our three knowledge worker participants from the accumulated data, and to consider what such an evaluation might say about the research question of how memory work contributes to the status and personal identity of technical communicators within their organizations.

At the time of this study, Angela still holds, if somewhat precariously, a level of mastery amounting to a hexis, or "an assured facility," in her work environment (Carruthers, 1990, p. 69). That is, she is still quite capable of bringing a significant degree of something equating to "maximum preparedness" to her work and to her dealings with the information development team and the other communities of practice with which she interacts. There was also very likely a span of time when Angela enjoyed a phronesis-level mastery that was recognized by virtually all of the employees at Software Unlimited. That was a time, as I asserted in chapter 6, when Angela's historical narrative was in very near lockstep with the overall memory regime of the company itself. Unfortunately, such is no longer the case, and Angela's state of phronesis no longer persists. And, as was evident from her confrontations with Carl, her state of hexis is now in danger as well.

In this light, Angela's litany-like list of the changes that Software Unlimited's expansion has precipitated, which I presented at the start of chapter 1, strikes us as a cry for help: "Now that we have six people on the
doc team . . . We also have a training department. . . . Now we've got processes. . . . There are a lot of meetings. We have a lot of projects and projects within . . . projects." Where, formerly, mastery consisted of and was
constituted by an ability to "just know," it now requires a different set of
practices, and Angela is finding herself slow to adapt: "There's not a great
way of keeping track of all that information at a glance, so you really rely
on remembering a lot."

Finally, to round out the set of inferences that might be made with respect to these information developers, the data show that Robert is quickly
developing hexis within his communities of practice. As well, he appears
to be on the fast track to one day achieving phronesis; a few more successful interventions with software developers in which he wins the day (such
as the one with Andie), will go a long way toward that goal. As for Lucy, it
would appear that she would be content, for now at least, with achieving a
hexis and that she is currently striving to do so, although her willingness,
even on a limited basis, to engage on a social plane with her community
of practice (e.g., arranging a problem-solving meeting with her colleague
Monica), shows promise toward one day gaining the self-confidence to pur-
sue phronesis.

MEMORY PRACTICES TRANSFORM INFORMATION
INTO KNOWLEDGE

While absolute mastery is probably unobtainable in knowledge workplaces—or is at least something of a moving target—Robert and Lucy show
us what technical communicators *can* do: they can leverage their rhetorical
memory practices as learning practices. As I noted in chapter 1, I employ
the terms "rhetorical memory practices" and "rhetorical memory work"
to describe activities that are elsewhere referred to as information- or
knowledge-managing activities in order to emphasize that it is the practices
themselves even more than the outcomes of those practices that transforms
information into knowledge. In other words, the activity of saving a piece
of information to an archive—a document into a file folder database, for
instance—does not merely preserve the piece of information for later use;
it transforms and changes the information in some way by consigning it
to some new order or enrolling it in some new endeavor. As Clark (2003,
p. 69) notes, "It just doesn't matter whether the data are stored somewhere

inside the biological organism or stored in the external world. What matters is how information is poised for retrieval and for immediate use as and when required." Even more importantly, the practice changes the practitioner who initiates it; it changes us. As I quoted Wenger (1998, pp. 56–57) earlier in chapter 3, "Participation in social communities shapes our experience [just as] it shapes those communities; the transformative potential goes both ways. . . . Our ability . . . to shape the practice of our communities is an important aspect of our experience of participation."

Further, this transformation, no matter how minor or apparently insignificant, occurs at multiple levels and across multiple members of the community: in the mind of the participant who initiates the practice (e.g., by repeatedly telling the story of her product's evolution, Angela gradually becomes a character in it), in their embodied dispositions (e.g., by repeatedly browsing her archive, Lucy cements knowledge of her product in both her mind and hand), and in the archive documents produced and circulated (i.e., communicated) through the practices themselves (Robert's reminder note takes on a life of its own during his meeting with the software developers).

MEMORY REGIMES BOTH CUE AND CONSTRAIN THE CIRCULATION OF INFORMATION

Similar to phronesis, a memory regime is best understood as an ongoing and evolving process rather than as a fixed product or stable condition. The regime can be perceived as a series of relational effects issuing from different hierarchical levels of the organization, or among communities of practice within the organization that lie essentially on the same plane of authority as each other. At its most formalized or "institutionalized" level, the memory regime consists of the infrastructures, tools, and conditions for success or failure (achieved via policies) that an organization as a whole puts in place. But an organizational memory regime also consists of less formal, sometimes unspoken policies or infrastructures—what Deuten and Rip (2000, p. 67) term "narrative infrastructures," which often hold sway at the level of communities-of-practice subcultures where, so often, the real work of the company gets done. So, at Software Unlimited, there exists—or at least, there appears to have existed for some time—an unwritten rule that knowledge of the backstories or developmental trajectories of its products is held in high regard and often represents the most persuasive warrant for arguments. This is a "rule" that is not without merit, because the underlying logic states something like "What has worked well for product creation and development in the past stands a reasonable chance, in principle, of working well in the future"—as long as due diligence, through user advocacy, for example—is accorded to both new technological innovation or new "bells and whistles," and to the constantly changing nature of consumer or customer demand. This is what underlies the adoption of a valid and useful memory practice: that it represents knowledge as an activity that has proven successful to getting a particular job done in the past and thus represents a valuable template for doing so in the future.

These infrastructures, tools, and policies induce a certain inertia, which enables us to discern something temporarily coherent enough to be labeled *the* memory regime of the organization, but they are also subject to change by influences coming from other spheres of influence within the company. This is especially true of organizations whose "raw materials" are information and knowledge itself, and for which the influx of new information and data is overwhelming and unrelenting it its volume, and fast and furious in its arrival, and perhaps no industry fits this description better or more aptly than the high-technology sector in which Software Unlimited orbits. This tsunami of information does not arrive at the company as through a funnel, landing first, let's say, with the company leaders in executive management, through which it can be "processed" into a coherent corporate theme before being handed down to managers and staff in the form of some corporate policy or the like. Rather, the information washes ashore, largely from the ubiquitous Internet, across all departments and even directly to individual employees of the company, where it is picked up—we might even say this happens somewhat randomly, because certainly that appears to be the nature of information and communication today—and transformed, utilized, reinterpreted, and communicated (or possibly not communicated) across different subcultural communities of practice (and other individuals) within the organization. Software Unlimited's rapid growth, combined with the independent spirit it espouses, both allows and encourages its employees—both teams and individuals—to aggressively seek out and exploit alternative methods of information gathering, as well as encouraging them to devise ingenious and even idiosyncratic ways of transforming that information into knowledge that will be useful, and profitable, for the company. These circumstances, as we have seen, can create what are, in effect, nested memory regimes in which each community of practice has its own preferred way of managing information and sharing knowledge—its own subcultural memory regime. So, again, at Software Unlimited, each product team has its own tools and methods for team knowledge sharing and each implements the Agile methodology in its own way and with its own preferred rigor.

As might be expected, a problem with this laissez-faire approach is that knowledge becomes fragmented and obscured, a source of contention among teams rather than a unifying factor. For instance, rather than being written down and stored on company file servers, the rationales or backstories behind product development decisions become the possessions of a few master practitioners, old-timers in specific communities of practice who are then permitted to share this tacit knowledge when, where, and with whom they see fit (it seems unlikely that Angela would have dismissed a junior information developer as brusquely and with as little explanation as she did the training team member).

Similarly, at Software Unlimited there exists no central or accessible repository of information about the company's users, causing each team to assemble its own collection of lore about users. These "silos" of information then become the source of turf wars as communities of practice attempt to safeguard their own stockpiles of hard-won information from other communities. Just such an incident was discussed in chapter 4, when the information development team was discourage from conducting its own usability tests and instead required to petition other teams for this information on an as-needed basis (Rockley, Kostur, & Manning, 2003). Another example can be discerned in chapter 6, in which the information development team feels compelled to come up with a new team name and mission statement that would better distinguish its user advocacy role from that of other, competing teams, particularly the UX team.

As I stated earlier, it is not within the scope of this book to evaluate the operational "health" of the company. Still, it is entirely relevant to point out that the crucial question facing Software Unlimited, the answer to which is critical to its continued success, is, how well does that "incoming" information get transformed into valuable knowledge *and also* effectively communicated through all levels of the organization? Hence, the critical importance of the work of the technical communicators in the Information Development Department who are the subjects of this study becomes obvious.

TOWARD AN ART OF MEMORY FOR THE TWENTY-FIRST-CENTURY TECHNICAL COMMUNICATOR

At last we return to the questions that motivate the study. The answer to the first question, concerning the type of information that is most important for technical communicators, initially seems obvious: technical communicators should know intimately and understand fully their company's products or services. This is true whether or not the technical communicator works directly with the product or not, but it is emphatically important if the technical communicator is expected to play a role in the design process, as well as in safeguarding to some extent the user experience and advocacy aspect, as the information developers at Software Unlimited are expected to do.

However, when we attempt to answer the second set of questions concerning where this information about a company's products and services lives and how it circulates, the situation appears far more complex. The full story of a product or service—of any designed artifact—must include the paths not taken, the disputes, the rationales, the bruised egos, and the rejected prototypes lying behind the artifact's current iteration. This information, as Software Unlimited shows, cannot be found on any spec sheet and often is not stored in any database.

This troubling answer to the "where" question leads us to consider the final questions: how do technical communicators locate and transform this obscure information into knowledge, and why should they? The answer to these questions provides intimations of what the art of memory for twenty-first-century technical communicators looks like.

I would suggest that this "new" art of memory actually bears a more-than-incidental resemblance to the ars memoria of the ancient and medieval rhetoricians. As I discussed in chapter 1, the object of rhetorical theory is similar to the object of much technical communication theory: it theorizes processes of invention or creation. Specifically, and again jibing with technical communication theory, rhetorical *memory* theories of the ars memoria focus primarily on how new things are created out of old things: in the case of the ancient rhetor, how he composes his speech by remembering elements of past speeches; in the case of the technical communicator, how she finds the information she needs to write a user guide by searching her archives, by talking to a subject matter expert, or by listening to an old-timer's stories.

In the ars memoria, the rhetor was able to find the right information at the right moment via the mechanism of tagging each bit information in memory with an image, and the "the chief features of a memory-image [were that it was] sensorily derived and emotionally charged. . . . Successful memory schemes all acknowledge the importance of tagging material emotionally as well as schematically, making each memory as much as possible into a personal occasion" (Carruthers, 1990, p. 59–60). In short, the techniques of the ars memoria, like the place-image mnemonic in which the orator imagined walking through a real or imaged place containing mnemonic images representing the information in his speech, worked so well because the memory images were both classified and rendered personally relevant (i.e., affectively, emotionally, physically significant) during the process of consigning them to the "database."

I would suggest in closing that the rhetorical memory practices that we have observed the information developers engaging in throughout the book—the finding, reminding, archiving, referencing, storytelling, and gesturing—all contribute to their memory images of their information. That is, the practices "tag" an otherwise formless piece of data by giving it some physical or emotional resonance that renders it usable in the future: Robert's archiving keeps his best ideas in the back of his mind; Lucy's browsing impresses the lineaments of her evolving product into her hand and mind, and Angela's storytelling invests her product and the history of its evolution with deep personal significance.

I began this book with the remarkable story of the truly gifted Derek Paravicini, who, despite blindness and severe autism, is nevertheless an accomplished musician and piano virtuoso. His talents, we might reasonably conclude, would seem to come about as close to the absolute perfection of hexis as any human being is apt to be able to achieve. And yet, with all due deference and with extreme admiration for Derek's miraculous abilities and talents, the information developers in this study are inundated, to the point of being overwhelmed, with far more in the way of cognitive, sensory, and physical inputs of information and data emanating from an ever-expanding multitude of sources—from email, texting, social media and all of the seemingly infinite Internet sites and sources, to more conventional sources of telecommunications, traditional media, and even snail mail and conversations with colleagues, managers, and outside professionals in their fields. Technical communicators must peer into the veritable "white noise" of the information age to assimilate all of this information, evaluate its merits— often evaluate its very legitimacy or veracity—and decide what information is useful and may turn out to be transformable into valuable knowledge for their organizations or companies, and discard or disregard information that is not useful, or that may actually be erroneous.

What they learn, the way that they learn it, and the way that they communicate this information-turned-knowledge to their colleagues and coworkers is through the individual memory practices that they employ, somewhat idiosyncratically, but also under the repertoire of acknowledged and accepted practices within their communities of practice. These memory practices are rhetorical in nature, and in large measure, information managers in the twenty-first century use the modern affordances of the digital workplace to achieve control over knowledge in much the same way that the ancient rhetoricians used mnemonic devices and tactics to memorize a speech.

APPENDIX A

This appendix provides a brief introduction to and company history of Software Unlimited and its assorted plant facilities. Next, it presents an introduction to the study participants along with a general description of their organizational relationships. The appendix next details the physical environs of the research sites, including infrastructures, workspaces, and pertinent company or office tools and equipment, both software and hardware. Finally, I conclude this appendix by providing my complete research methodology for this case study, including all of the following aspects: data collection, data preparation, and data segmentation for analysis.

OVERVIEW OF SOFTWARE UNLIMITED

Software Unlimited is a medium-sized, privately owned software company located in the midwestern United States that creates and markets graphics software to consumer, corporate, and university customers throughout the United States and in thirty other countries. As part of its increasing focus on international expansion, the company's products, including software and documentation, are made available in five languages. At the time of the study, Software Unlimited had been in existence for almost twenty years, during which time the company had experienced periods of rapid growth as well as periods of downsizing. In the two years immediately prior to the study, the company had undergone tremendous growth, nearly tripling its number of employees, to about two hundred.

The vast majority of the employees are based in Software Unlimited's headquarters offices, which are housed in four separate buildings leased within a small suburban office park. This fragmentation of office spaces being an unplanned byproduct of the company's rapid expansion, the company had ambitious plans to build a new state-of-the-art headquarters facility in order to consolidate its employees in one building and to provide them with office layouts and furnishings more conducive to team-based collaborative work. As part of these plans, during the period of the study, the company was experimenting with new office configurations and types of furnishings in one wing of one suite of offices, which was a factor in the research described in this book. The majority of the research data reported in this study were gathered in the building housing the technical communicators as well as the company executives, administrative staff, and a significant portion of the software developers.

PARTICIPANTS: THE INFORMATION DEVELOPMENT TEAM AND THE RESEARCHER

This case study focuses on the five technical communicators and their manager who made up the information development team at Software Unlimited. The information developers reported to the information development manager for general oversight and quarterly reviews. In addition to being housed in proximity to each other, the team met weekly on Wednesday afternoons. Each information developer was also assigned to a separate Software Unlimited software product and served on that product's development team along with software developers, quality assurance engineers, and representatives from the Training Department. The information developers attended daily and project milestone meetings with their product teams and participated in marketing events related to their product.

The research participants consisted of four women and two men, each with varying degrees of experience as technical communicators and different lengths of tenure with the company, ranging from a few months to almost six years. The writers chose or were assigned the pseudonyms Angela, Peter, Robert, Monica, Lucy, and Becky, the team manager. Table A.1 offers a brief introduction to the entire team.

In addition to the participants, the researcher was always present during the six observation sessions reported in this book and his presence inevitably influenced the action of the scene in some way whether this presence was immediate and physical or remote and delegated, such as when a video or screen recorder was employed to capture data (Sullivan & Porter, 1997; Tedlock, 2003). A final participant in all observation sessions, therefore, was the researcher himself. Consequently, although I tried to minimize my influence on the other actors, I include myself in accounts of observations by noting my physical (or remote) position in the accounts and by noting

TABLE A.I. The Information Development Team

Becky	Information development team manager. Worked for Software Unlimited as an information developer for about six months before being promoted to team manager. Has worked in technical and professional writing for fifteen years and holds a bachelor's degree in professional writing.
Angela	Most senior member of the information development team. Was the first information developer ever hired at Software Unlimited and for several years the only one. Has now worked at the company for almost six years. Before that, was a process improvement engineer and a consultant. Had returned to college in midlife and earned a bachelor's degree in technical writing.
Robert	Newest member of the information development team. Has worked at Software Unlimited for just over two months. Before that, worked as a freelance writer for several high-tech companies and published two books on technology. Holds a bachelor's degree in elementary education and a master's degree in educational technology.
Monica	Has worked at Software Unlimited for about two years, her first job out of college. Holds a bachelor's degree in professional writing.
Lucy	Has worked for Software Unlimited just over two years. Was on maternity leave and part-time work for one year during this period. Returned to full-time status about one month before the initial interview for this study. Holds a bachelor's degree in technical communication.
Peter	Has worked at Software Unlimited for about ten months, his first job out of college. Holds bachelor's degrees in telecommunications and English.

what perceivable influence my presence may have had on the other participants, and, where relevant, by speculating about how my influence could have been minimized.

INSIDE THE RESEARCH SITES: INFRASTRUCTURES, WORKSPACES, AND TOOLS

The information development team was located in a single suite of offices, which they shared with several other groups of employees (see fig. A.1). Peter, Robert, and Lucy were assigned to cubicles in the center of a large room in one wing of this suite, and Monica, Angela, and Becky occupied offices around the perimeter of this room. Other company employees (i.e., employees who were not information developers) occupied the remainder of the cubicles and offices in this room. One end of the large room was devoid of cubicles and left open in order to be used exclusively by the Graphic Forge product team, to which Angela belonged.[1] This open area and the of-

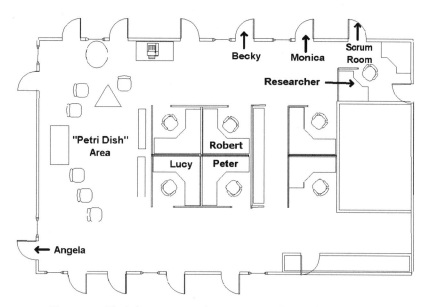

Figure A.1   The Information Development Wing. The Graphic Forge team
is also housed in this wing.

fices immediately surrounding it were nicknamed the Petri Dish, because
the area contained experimental furnishings and office configuration for the
planned new corporate headquarters. Because it was part of the experiment,
only the Graphic Forge team was permitted to hold meetings in this area.
The furnishings in the open area, including tables, chairs, whiteboards, and
a video projection screen, were all wheeled so that they could be reconfig-
ured as needed. The offices at this end of the room each housed two employees' workstations.

Within these offices, these employees' workstations were placed in
several experimental configurations—some occupants' desks faced each
other, others faced away from each other, and still others faced the same
direction. Two other information development team members, Monica and
Becky, the manager, occupied conventionally arranged and furnished offices
at the other end of the room. Robert's, Peter's, and Lucy's cubicle walls were
five feet tall, giving them some physical privacy but not shielding them
from office noise.

At the other end of the large room were two meeting spaces: the library
and a small meeting room termed the Scrum Room, for reasons that will be
explained. The library held a large conference table with seating room for

ten, a projection screen, and bookshelves along the perimeter, filled with books and periodicals of importance to the software industry, as well as video games and movies which could be checked out by employees. Although some differences in the writers' workspaces existed and will be noted later where significant, each workspace possessed, at minimum, a desk, chair, a small file cabinet, a laptop or desktop computer, and at least one monitor. With a few exceptions, which I will note later, most of the data collection reported in this book took place in the spaces described above.

As would be expected of a software company, the computer hardware and software infrastructures supporting the information developers were extensive, and participants' configurations and use of these infrastructures varied. However, a few commonalities across participants can be given here. Every participant had at least one personal computer in his or her workspace, though several participants had more than one, including a mix of laptops and desktops and Macs and PCs. Every participant had at least one external monitor and keyboard, though, again, several had more than one. Every participant had the Microsoft Office suite installed on his or her computer, as well as a suite of graphics-editing software. Every participant used Microsoft Messenger for instant messaging and Microsoft Outlook for email. Additionally, several participants used the Outlook calendar feature for work scheduling and reminders. Every participant had access to the Internet, a shared network drive, a shared company intranet, and a shared company wiki.

A few other aspects of the infrastructure of Software Unlimited are worth noting. First, the firm's modest size meant that there was no team of IT professionals exclusively dedicated to centrally managing the company's databases or other memories, which created a situation similar to the prototypical one I outline in chapter 1 in which technical communicators may be expected to become the de facto information brokers for their teams (the title "information developer" itself reflected this). Second, Software Unlimited's explosive recent expansion created a somewhat haphazard and provisional set of workspaces and infrastructures characteristic of companies in a "fast capitalist" economy (Gee, Hull, & Lankshear, 1996). Finally, Software Unlimited's longevity (for a software company) combined with its rapid growth made it a particularly good site in which to observe a range of issues related to innovative versus entrenched ways of doing work. In sum, Software Unlimited appeared to be both a relatively typical but also especially revealing place to study the kinds of knowledge-intensive memory work in which technical communicators now increasingly engage.

DATA COLLECTION

Studying memory practices and the memory regimes that arise from them is a complex undertaking requiring a great deal of fieldwork and a mix of research and data collection methods. For the present study, I spent over 120 hours at Software Unlimited during a six-month period (September 2007—March 2008) engaged in a range of ethnographic fieldwork activities. These activities included conducting extended initial interviews with each research participant, job shadowing, observing composing sessions, attending meetings, and holding short follow-up interviews whenever possible after research sessions. I collected data during these activities using a variety of tools and methods, depending on the needs of the situation and the preferences of my participants. These data collection methods included field journaling, video recording, screen capturing, and collecting artifacts (i.e., digital texts or paper documents that my participants were kind enough to share with me). Ten work sessions, including four observation sessions and six initial interviews with the information development team members, were video recorded, resulting in approximately fifteen hours of video.

Beginning my first week at the research site and continuing over the course of the entire six months of the study, I attended the weekly information development team meetings, sat in on meetings between the information developers and members of other teams, and spent entire working days listening to and watching my participants from a cubicle that Software Unlimited had provided for me adjacent to the information developers' workspaces.

Beginning at the end of the first month of the study, I interviewed each of the six information developers to learn about his or her background (including educational background, history with the company, and experience as a technical communicator) and memory and writing tools and practices. These interviews were semistructured (Prior, 2004), consisting of a core set of questions (see app. B) posed to participants in an open-ended manner by adopting a conversational tone, by asking unscripted follow-up questions where I thought more information would be helpful, by omitting certain questions if I thought they had already been sufficiently answered, and by interjecting comments or affirmations where I felt they were expected.

I also audio and video recorded these interviews using a Webcam attached to a laptop computer. The Webcam was focused on the participants for the duration of each interview in order to record participants' gestures and movement as they answered the interview questions. The purpose of videotaping the interviews was to provide additional data sources in order to create a richer understanding of the unfolding dialogue.

Additionally, thirty work sessions, meetings, and impromptu interviews were recorded in researcher field journals, resulting in approximately ninety pages of handwritten notes. Last, three of the video-recorded work sessions were also simultaneously screen recorded.

Together, these activities and methods enabled me to assemble accounts of scenes with enough durability and specificity to answer the research questions I articulate at the end of chapter 1 concerning the relationship between the memory regime at Software Unlimited and the memory practices of its technical communicators.

DATA PREPARATION

These research methods produced four basic categories of data: (1) handwritten field notes, (2) video recordings, (3) computer screen recordings, and (4) artifacts such as printed and electronic documents produced by the research participants. Each of these data types required different steps in order to prepare it for segmentation and coding. For handwritten field notes, as soon as possible after an observation session (typically within twenty-four hours), I converted my notes into coherent write-ups and attached to these write-ups cover page contact summary forms (Miles & Huberman, 1994), on which I recorded salient contextual information about the session, including the participants involved, the location, any software or hardware employed by the participants, and any particular issues, themes, or unanswered questions related to my research questions that arose during the scene.

Computer screen recordings were made using TechSmith Morae, a usability testing research software package that Software Unlimited employs in its own user testing efforts. For each of the Morae recordings, I transcribed the audio and made extensive annotations from the videos, noting participant gestures, facial expressions, posture, and object manipulations indexed to the corresponding audio. Two work sessions were recorded with Morae capturing onscreen actions (cursor movements, clicks, and keyboard activities) indexed to audio and video recorded via a webcam located atop the participant's monitor. For these recordings, I transcribed speech and wrote detailed accounts of participant computer activities and any gestures or physical object manipulations captured on the video. Additionally, as with the work sessions recorded using field notes, as soon as possible after the

observation session, I created cover page contact summary forms on which
I recorded background information about the session.

DATA SEGMENTATION:
D-UNITS IN TRANSCRIPTIONS OF SPOKEN TEXT

During the data-coding phase of the project, I coded interview and meeting
transcripts into discourse units or "d-units" (Colomb & Williams, 1985).
Colomb and Williams (1985, p. 102) define a d-unit as "any stretch of continuous text—a whole text, a section, a paragraph, even a small group of related sentences—that functions as a unit and whose parts are more related
to each other than to those outside the d-unit." The central structure of a
d-unit is an issue statement and a discussion about that issue. Thus formulated, the d-unit proves to be a particularly appropriate tool for rhetorical
analyses because it enables the researcher to disaggregate texts at a middle,
or mesoscopic, level. That is, where microscopic units of analysis like the
t-unit are of little value in helping segregate portions of text according to
rhetorical function, and larger groupings like conversational exchanges and
turns can elide or obscure important but unexpected occurrences like slips,
asides, or interjections that occur regularly in spoken language, d-units are
scalable and nestable on the basis of the differences the reader, interlocutor,
or researcher perceives between them.

    First, the d-unit is scalable because the component parts, the issue and
discussion, need not correspond to sentences or even groups of sentences
but instead are "fixed discourse position[s] or slot[s]" that can vary in
length, ranging from a clause to a paragraph, depending on the phenomena
of interest to the researcher (Colomb & Williams, 1985, p. 108). This proved
especially helpful for a study with an interest in the operations of memory,
because it enabled me to code as separate the various meandering asides and
self-interruptions that the associative nature of human memory makes inevitable when we talk about events of the past as we do during interviews.
The d-unit, thus, represents a finer level of segmentation than the turn or
the exchange. To give an example, when responding to the interview question "What resources do you typically consult when conducting . . . research
[for a new project]?" the information developer Robert began by describing
how he often employs coworkers as resources:

> I try to make friends with all the programmers who are doing the website
> or doing the recorder or anything. Now it's actually getting better—on
> Monday morning or Tuesday—[Becky] and I had a meeting with some

marketing people, which doesn't really happen that often. So I'm learning how to get something done. So today for example, there's some technical things that I don't really understand or spend my time on like how
to download or install or point my recorder to the right server. So now I
know this . . . guy who'll take care of that problem really quick for me.

But, as Robert spoke, he began to describe a different sort of resource:

Other things, [Software Unlimited] on the N drive has a history of all
the things associated with a project. So if I went into [product name]
when I got here I could kind of go back and look at what had been said
in other meetings and Sprints. So far it seems like it's not really organized or used very well, so one of my goals with the wiki would be that
I could click on [product name] and there'd be a lot better organization
of what's going on.

The flexibility of the d-unit enabled me to segment these differing answers
to the same interview question separately.

A second affordance of d-units is that d-units can be nested inside other
d-units, depending on the phenomena of interest and the needs of the researcher. This characteristic makes the d-unit powerful for coding interviews and for accounting for speech *about* an activity as *an indicator of*
that activity. In the case of interviews, particularly questions that ask participants to search their long-term memories for complex answers, the concept of nested d-units enables the researcher to segment the inevitable side
roads and digressions that human associative memory produces. In the case
of activities, the d-unit can also help index talk about activity to the activity such talk describes. For example, continuing the example above, after
talking about nonhuman resources for a while, Robert seamlessly resumed
talking about his reliance on his colleagues: "If I got transferred to [another
product], that would be really scary to me because I don't know much about
it. . . . So the first thing I would do is talk to Peter, who's on it now, and get
the behind the scenes kind of [stuff]." The d-unit enables Robert's discussion of his memory practice of searching the network hard drive to be lifted
out and analyzed separately from the two parts of his discussion of consulting his colleagues, which bracket it.

Initial Interview Questions

Questions about Participant Background

1. What is your current job title?
2. What are your job duties?
3. How long have you worked as a technical writer?
4. How long have you worked as a technical writer at Software Unlimited?

Questions about Composing and Writing Practices

1. When you are assigned to a project, what sorts of research do you perform in order to prepare yourself to work on that project? Some examples of this type of activity might include reading the project design documents or meeting informally with other members of the project team to discuss the project.
2. What resources do you typically consult when conducting this research? For example, do you search for information on the company file server or the Internet, do you consult colleagues?
3. How do you track and organize the information you learned through this research? For example, how and where do you keep notes and other products of your research?
4. Where do you typically compose?
5. When do you typically compose?
6. How do you prepare your workspace (both your computer desktop and your physical surroundings) before beginning a writing task? Examples of this type of activity might include creating computer file folders for the writing deliverables or creating a physical file folder for hardcopies of project documents.
7. What software tools do you use when composing?

8. In addition to the project deliverables, do you create supplementary documents or files for yourself to help you manage information or track your goals as you compose? An example of a supplementary document might be a working outline to remind you of your place in the composing process.

9. Do you use any supplementary software tools to support your composing? For example, you might use a text editor like Windows Notepad to create quick reminders for yourself as you write.

10. Once you have begun writing, do you take many breaks or do you work straight through until you come to a logical stopping point? If you take breaks, what activities do you typically perform during them? For example, do you talk to colleagues, visit websites, or go to the break room?

11. How often do you communicate with your colleagues about the project, and what methods do you use for this communication? For example, do you meet with them face-to-face, do you instant-message them, do you email them?

Questions about Project Management Practices

1. What methods and tools does your employer use to help you track project due dates? For example, what software (e.g., MS Project) or hardware (e.g., a daily planner) do you use?

2. Do you have any personal methods for managing projects that supplement those your employer provides? For example, do you post sticky notes for yourself? If so, is there a system behind these personal methods (e.g., placing the most important sticky note on your monitor and the rest on your bulletin board)?

3. Do you find that, as a project continues, the project management aspects of the writing process grow easier or more difficult? In other words, does managing the various deliverables and documents you and others create for a project become more or less time consuming as the project continues? Please explain why.

NOTES

CHAPTER I

1. I use pseudonyms for the company and for all of the participants. My research was approved by the Institutional Review Board for Human Subjects Research at Michigan State University.

2. In fact, the earliest chronicle of early modern office life that has come down to us, Samuel Pepys's diary, recounts conditions of distraction and fragmentation that will be familiar to anyone who has ever had to "grin and bear it" at the office despite pressing personal exigencies. As Clair Tomalin (2002, p. xxxi) describes it, "Pepys lets us know that each of us inhabits a perpetually fluctuating environment, and that we are changed, moved and sometimes controlled by our inner tides and weather fronts even when we are most engaged in official functions."

3. Social constructionist approaches to memory often reject or, at best, regard as "misdirected and mistaken" cognitive theories of memory and laboratory-based research methods (Shotter, 1990, p. 126). The principal rationales in social constructionism for rejecting attempts to explain the functioning of memory in terms of interior constructs are (1) that there is no such thing as an autonomous individual mind, because all thinking (and remembering) is inevitably entangled in social experience; and (2) even if such autonomous thought and memory were possible, there would be no way to verify the accuracy of such constructs without recourse to social activities anyway (e.g., talking about memory): "The correctness of an inner process cannot be tested by comparison with yet another inner process. At some point, reference to activities in daily life at large is necessary" (Shotter, 1990, p. 127).

It seems to me, however, that too quickly drawing a rigid line between cognitive and social explorations of memory cuts off enlightening and potentially useful explanations for the multifarious phenomena of memory that we witness in both daily life and situated research. In fact, social constructivist research approaches like situated-cognition theory (on which a great deal of the current study is based) already bridges this gap by exploring the mechanism by which the external, social world gets inside our heads. This mechanism is, of course, our body, or, more specifically, our embodied sensorium, which, as Aristotle noted nearly twenty-five hundred years ago, is the origination point for most of our memories: "We must conceive that which is generated through sense-perception in the sentient soul [as] the state whereof we call memory" (Aristotle, 1952, p. 691). Thus, although my approach to studying memory is fundamentally social, I do occasionally discuss the overlaps that I perceived to exist between social and cognitive, occasionally even neuronal, research.

4. Even recent technologies, such those that enable cloud computing, adhere to metaphors several generations removed from current work paradigms. For example, Dropbox, one of the most successful current cloud computing platforms, employs the file folder as its organizational metaphor.

CHAPTER 2

1. Note here that I do not wish to imply that such internal competition is necessarily healthy to the long-term prospects of the firm. Indeed, the preponderance of the literature (e.g., Kanter, 1983; McLean, 2005) demonstrates that internal competition among work teams and communities of practice more often than not leads to unhealthy siloing of knowledge within firms—an effect reaffirmed by the present study. Rather, at this point at least, I wish merely to call attention to the rhetorical and even occasionally agonistic role that memory practices may play for the individual worker.

2. In fact, as I noted earlier, Vygotskian concepts of activity theory like the notion of tool mediation, with its focus on external physical actions as keys to understanding internal mental states and its assertion that "perception is an integral part of human interaction with the world" have gone a long way toward refuting the legacy of post-Cartesian understandings of human thought as a product of an asocial, ahistorical, a-material, and disembodied mind and, consequently, toward revalidating Aristotle and his successors' embodied epistemology (Kaptelinin & Nardi, 2006, p. 81).

CHAPTER 4

1. By identifying these cross-functional, Agile-inspired product development teams as communities of practice, I am employing the concept somewhat flexibly but in a way that is, nevertheless, consistent with Wenger's definition. That is, although Wenger (1998) cautions against treating the concept of a community of practice as "a synonym for a group, team, or network" (p. 74), he goes on to define communities of practice as, most fundamentally, "shared histories of learning" (p. 86). So, while we cannot assume a priori that flexible, cross-disciplinary work groups like the product teams at Software Unlimited become communities of practice at the moment those groups are assembled by management fiat, we can assert with a good degree of certainty that they will have become so by the end of their first thirty-day Sprint. In fact, it could be argued that a central purpose of the Agile methodology, with its emphasis on consistent Daily Scrums bracketed by rigorous planning and review meetings, is to generate the "shared histories of learning" that quickly transform loose assemblages of individual workers into focused communities of practice.

2. Recall that, as I distinguish them, archiving practices are attempts to preserve information in memory for longer periods of time and for multiple or less clearly demarked

purposes, while reminding practices are performed chiefly to preserve information for a limited time and for a single (or very limited number) of uses.

3. Only Angela, the information developer with the longest tenure with the company, differs considerably in her approach to memory work. Understanding this difference provides an important clue to understanding the memory regime at Software Unlimited. Angela's case is discussed in detail in chapter 6.

CHAPTER 5

1. When asked if she shared these data dump documents with members of her product team, Lucy replied in the negative: "Um, I put it out there—at first I was telling them where they were, but nobody was looking at them. So I haven't done that [i.e., told the team about the documents]." It would seem, unfortunately, that Lucy's "data dumps" are of little use to anyone else in the organization.

2. In an activity theory framework, browsing could be thought of as semiautomatic, semiconscious *operation* rather than a conscious, deliberate *action*.

3. Recall that in chapter 4, a similar change to his product's interface prompted Robert to begin writing his reminder note to take to his Daily Scrum. He could tell that the product interface and functionality had changed with the most recent build, but he was not entirely sure how it had changed or if it would change again without his intervention.

4. In this interview transcript, I have included descriptions of Lucy's hand gestures and bodily movements in brackets and indexed these movements to individual frames extracted from the video recording of the interview. A diagram illustrating these movements indexed to Lucy's speech is presented in figure 5.5.

5. The fact that the contents of the Sprint directory depicted in figure 5.4 and Lucy's memory image of the database depicted in figure 5.5 do not match exactly (e.g., Lucy remembers "Camera Recording" but Sprint 21003 contains a directory titled "Two_Camera") does not negate the mnemonic benefits of browsing: Lucy's memory may not be perfect but it does appear stronger than it would have been otherwise.

CHAPTER 6

1. No explanation was offered for why Angela herself did not continue in the position of manager. However, the purpose of this study is to examine the memory and information management practices of technical communicators and how those practices may be understood as components of a rhetorical art of memory. As a result, it is beyond the scope of this study to make judgments regarding the existing management hierarchy at Software Unlimited. Nevertheless, the findings of this study may offer some indicators underlying the practical basis for that hierarchy as it exists. As an aside, however, whatever those indicators might show, Angela deserves a great deal of credit for her role in helping to grow the company. User satisfaction is particularly critical to the success or failure of software companies. That fact that Angela appears to have been virtually single-handedly responsible for the process improvements that made Software Unlimited's products so user friendly as to spawn such prodigious growth says a great deal about the extreme value of her contribution to the organization.

2. Graphic Forge is the graphics-manipulating software product to which Angela is assigned. The name has been changed to protect confidentiality.

3. This excerpt from my initial interview with Angela appears somewhat different from the other memory work discussions recounted in this book because it is dominated by one actor's verbal account of past action. However, I selected this excerpt for extended analysis for the following reasons: (1) the story that Angela tells here is central to her identity at Software Unlimited; (2) storytelling is a key memory practice (perhaps *the* key memory practice) by which Angela translates past information to meet present exigencies (i.e., how she meets the needs of kairos); and (3) this story was retold or referenced, either in part or in its entirety, multiple times throughout the study, and it influenced (i.e., acted upon) other individuals within the organization, most profoundly the other two participants in this case study, Robert and Lucy, across a variety of other significant contexts. In short, the story served as an indicator of Angela's phronesis.

4. Ironically, although her efforts proved unsuccessful with the initial release, in the end, Angela was proven correct: six months after the initial release, in response to customer complaints, a major upgrade was released that fixed many of the issues the information developers identified in their December 5 meeting. However, with a different issue guided perhaps by a different set of facts or circumstances, the outcome could easily go the other way. The important point is that when different subcultures adhering to different memory regimes exist within a single organization, there is the possibility—even the probability—of competition and conflict among the intraorganizational memory regimes of different communities of practice.

CHAPTER 7

1. Although each information developer simultaneously participates in multiple communities of practice, including the information development team itself, it is to their product development teams that the majority of their time and work efforts and the preponderance of their concerns with team relationships were directed.

APPENDIX A

1. Pseudonyms for Software Unlimited products have been used in order to preserve confidentiality.

REFERENCES

Ackerman, M. S., & C. Halverson. 1998. Considering an organization's memory. *Proceedings of the 1998 ACM Conference on Computer Supported Cooperative Work*, Seattle, WA, November 1998, 39–48.

———. 2000. Reexamining organizational memory. *Communications of the ACM* 43 (1): 59–64.

Albers, M. J. 2005. The future of technical communication: Introduction to this special issue. *Technical Communication* 52 (3): 267–272.

Allen, V. 1993. The faculty of memory. In *Rhetorical memory and delivery: Classical concepts for contemporary composition and communication*, edited by J. F. Reynolds, 45–63. Hillsdale, NJ: Lawrence Erlbaum Associates.

American Association of University Women. 2003. *Women at work*. Washington, DC: American Association of University Women Educational Foundation.

Amidon, S., & S. Blythe. 2008. Wrestling with proteus: Tales of communication managers in a changing economy. *JBTC* 22 (1): 5–37.

Anderson, D. L. 2004. The textualizing functions of writing for organizational change. *Journal of Business and Technical Communication* 18 (2): 141–164.

Aristotle. 1952. On memory and reminiscence. In *The works of Aristotle*, vol. 1, translated by W. D. Ross, Great Books of the Western World, edited by R. M. Hutchins, vol. 8, 690–695. Chicago: Encyclopaedia Britannica.

Bannon, L. J., & K. Kuutti. 1996. Shifting perspectives on organizational memory: From storage to active remembering. *Proceedings of the 1996 Hawaii International Conference on System Science*, Wailea, HI, January 1996, 156–167.

Barab, S. A., &, J. A. Plucker. 2002. Smart people or smart contexts? Cognition, ability, and talent development in an age of situated approaches to knowing and learning. *Educational Psychologist* 37 (3): 165–182.

Barker, T., & K. Poe. 2002. The changing world of the independent: A broader perspective. *Technical Communication* 49 (2): 151–153.

Barnes, C. A., & B. L. McNaughton. 1985. Spatial information: How and where is it stored? In *Memory systems of the brain*, edited by N. M. Weinberger et al., 49–61. New York: Guilford Press, 1985.

Barreau, D., & B. Nardi. 1995. Finding and reminding: File organization from the desktop. *SIGCHI Bulletin* 27 (3): 39–43.

Bartlett, F. C. (1932) 1964. *Remembering: A study in experimental and social psychology.* Reprint, London: Cambridge UP.

Bellazza, F. A., & D. K. Buck. 1988. Expert knowledge as mnemonic cues. *Applied Cognitive Psychology* 2:147–162.

Besser, H. 2002. The next stage: Moving from isolated digital collections to interoperable digital libraries. *First Monday* 7 (6, June 3). Retrieved January 17, 2007, from http://www.firstmonday.org/issues/issue7_6/besser/index.html.

Bizjak, P. 2000. Mankind's memory managers: A new paradigm of library science. *Library Philosophy and Practice* 2 (2). Retrieved January 17, 2007, from http://www.webpages.uidaho.edu/~mbolin/bizjak.html.

Blackler, F., & S. McDonald. 2000. Power, mastery and organizational learning. *Journal of Management Studies* 37 (6): 833–851.

Blair, A. 2004. Note taking as an art of transmission. *Critical Inquiry* 31 (Autumn): 85–107.

Blakeslee, A. M. 1997. Activity, context, interaction, and authority: Learning to write scientific papers in situ. *Journal of Business and Technical Communication* 11 (2): 125–169.

Bourdieu, P. 1990. *The logic of practice.* Translated by R. Nice. Stanford, CA: Stanford UP.

Bowker, G. C. 2005. *Memory practices in the sciences.* Cambridge, MA: MIT Press.

Bowker, G. C., & S. L. Star. 1999. *Sorting things out: Classification and its consequences.* Cambridge, MA: MIT Press.

Brown, J. S., A. Collins, & P. Duguid. 1989. Situated cognition and the culture of learning. *Educational Researcher* 18 (1): 32–42.

Brown, J. S., & P. Duguid, P. 1991. Organizational learning and communities-of-practice: Toward a unified view of working, learning, and innovation. *Organizational Science* 2 (1): 40–57.

———. 2002. *The social life of information.* 2nd ed. Boston: Harvard Business School Press.

Buchanan, R. 1989. Declaration by design: Rhetoric, argument, and demonstration in design practice. In *Design discourse: History, theory, criticism,* edited by Victor Margolin, 91–109. Chicago: University of Chicago Press. (Original work published 1985.)

Carruthers, M. J. 1990. *The book of memory: A study of memory in medieval culture.* New York: Cambridge UP.

———. 1998. *The craft of thought: Meditation, rhetoric, and the making of images, 400–1200.* Cambridge: Cambridge UP.

[Cicero]. 1954. *Rhetorica ad Herennium.* Translated by H. Caplan. Cambridge, MA: Harvard UP for the Loeb Classical Library.

Clancey, W. J. 1997. *Situated cognition: On human knowledge and computer representations.* New York: Cambridge UP.

Clark, A. 2003. *Natural-born cyborgs: Minds, technologies, and the future of human intelligence.* New York: Oxford UP.

Colomb, G. G., & J. M. Williams. 1985. Perceiving structure in professional prose: A multiply determined experience. In *Writing in nonacademic settings,* edited by L. Odell & D. Goswami, 87–128. New York: Guilford Press.

Conklin, J. 2007. From the structure of text to the dynamics of teams: The changing nature of technical communication practice. *Technical Communication* 54 (2): 210–231.

Cooper, L. A., & J. M. Lang. 1996. Imagery and visual-spatial representations. In *Memory: Handbook of perception and cognition*, 2nd ed., edited by E. L. Bjork & R. A. Bjork, 129–64. San Diego: Academic Press.

Cornoldi, C., & R. De Beni. 1991. Memory for discourse: Loci mnemonics and the oral presentation effect. *Applied Cognitive Psychology* 5:511–518.

Crowley, S. 1993. Modern rhetoric and memory. In *Rhetorical memory and delivery: Classical concepts for contemporary composition and communication*, edited by J. F. Reynolds, 31–44. Hillsdale, NJ: Lawrence Erlbaum Associates.

Crowley, S., & D. Hawhee. 2012. *Ancient rhetorics for contemporary students.* 5th ed. Boston: Pearson.

Dalton, B. 2004. Creativity, habit, and the social products of creative action: Revising Joas, incorporating Bourdieu. *Sociological Theory* 22 (4): 603–622.

Davenport, E., & H. Hall. 2002. Organizational knowledge and communities of practice. *Annual Review of Information Science and Technology* 36:170–227.

Davis, M. T. 2001. Shaping the future of our profession. *Technical Communication* 48 (2): 139–144.

de Certeau, M. 1984. *The practice of everyday life.* Berkeley: University of California Press.

Delmas, B. 2001. Archival science facing the information society. *Archival Science* 1:25–37.

DeLong, D. W. 2004. Lost knowledge: Confronting the threat of an aging workforce. New York: Oxford UP.

Derrida, J. 1998. *Archive fever: A Freudian impression.* Translated by E. Prenowitz. Chicago: University of Chicago Press. (Original work published 1995.)

Deuten, J., & A. Rip. 2000. The narrative shaping of a product creation process. In *Contested futures: A sociology of prospective techno-science*, edited by N. Brown, B. Rappert, & A. Webster, 65–86. Burlington, VT: Ashgate.

Dicks, R. S. 2010. The effects of digital literacy on the nature of technical communication work. In *Digital literacy for technical communicators: 21st century theory and practice*, edited by R. Spilka, 51–81. New York: Routledge.

Dilger, B. 2006. Extreme usability and technical communication. In *Critical power tools: Technical communication and cultural studies*, edited by J. B. Scott, B. Longo, & K. V. Wills, 47–69. Albany: State University of New York Press.

Duguid, P. 2005. "The art of knowing": Social and tacit dimensions of knowledge and the limits of the community of practice. *Information Society* 21:109–118.

Dunne, J. 1993. *Back to the rough ground: "Phronesis" and "techne" in Modern Philosophy and in Aristotle.* Notre Dame, IN: University of Notre Dame Press.

Edmunds, A, & A. Morris. 2000. The problem of information overload in business organisations: A review of the literature. *International Journal of Information Management* 20:17–28.

Engeström, Y. 1999. Activity theory and individual and social transformation. In *Perspectives on activity theory*, edited by Y. Engeström, R. Miettinen, & R.-L. Punamäki, 19–38. New York: Cambridge UP.

Engeström Y., & D. Middleton. 1996. Introduction: Studying work as mindful practice.

In *Cognition and communication at work*, edited by Y. Engeström & D. Middleton, 1–14. New York: Cambridge UP.

Fentress, J., & C. Wickham. 1992. *Social memory*. Oxford UK: Blackwell.

Fertig, S., E. Freeman, & D. Gelernter. 1996. "Finding and reminding" reconsidered. *SIGCHI Bulletin* 28 (1): 66–69.

Flower, L., & J. R. Hayes. 1981. A cognitive process theory of writing. *College Composition and Communication* 32 (4): 365–387.

Fox, S. 2000. Communities of practice, Foucault, and actor-network theory. *Journal of Management Studies* 37 (6): 853–867.

Freeman, E., & D. Gelernter. 2007. Beyond lifestreams: The inevitable demise of the desktop metaphor. In *Beyond the desktop metaphor: Designing integrated digital work environments*, edited by V. Kaptelinin & M. Czerwinski, 19–48. Cambridge, MA: MIT Press.

Fussell, S. R., R. E. Kraut, & J. Siegel. 2000. Coordination of communication: Effects of shared visual context on collaborative work. *Proceedings of CSCW 2000*, December 2–6, 2000, Philadelphia, PA, 21–30.

Gabriel, Y. 2000. *Storytelling in organizations: Facts, fictions, and fantasies*. New York: Oxford UP.

Gee, J. P. 1990. *Social linguistics and literacies: Ideology in discourses*. London: Falmer Press.

Gee, J. P., G. Hull, & C. Lankshear. 1996. *The new work order: Behind the language of the new capitalism*. Sydney: Westview Press.

Geisler, C. 2001. Textual objects: Accounting for the role of texts in the everyday life of complex organizations. *Written Communication* 18 (3): 296–325.

Gelbard-Sagiv, H., R. Mukanel, M. Harel, R. Malach, & I. Fried. 2008. Internally generated reactivation of single neurons in human hippocampus during free recall. *Science* 322 (5898): 96–101. Retrieved February 5, 2010, from Academic Search Premier database.

Giammona, B. 2004. The future of technical communication: How innovation, technology, information management, and other forces are shaping the future of the profession. *Technical Communication* 51 (3): 349–366.

Goodenough, W. H. 1971. *Culture, language and society*. McCaleb Module in Anthropology. Reading, MA: Addison-Wesley.

Goodwin, C., & M. H. Goodwin. 1996. Seeing as a situated activity: Formulating planes. In *Cognition and communication at work*, edited by Y. Engeström & D. Middleton, 61–95. New York: Cambridge UP.

Grayling, T. 1998. Fear and loathing of the help menu: A usability test of online help. *Technical Communication* 45 (2): 168–179.

Hall, S. 1996. On postmodernism and articulation: An interview with Stuart Hall. In *Stuart Hall: Critical dialogues in cultural studies*, edited by L. Grossberg & K. Chen, 131–150. New York: Routledge.

Hallett, T. 2003. Symbolic power and organizational culture. *Sociological Theory* 21 (2): 128–149.

Hart, H., & J. Conklin. 2006. Toward a meaningful model for technical communication. *Technical Communication* 53 (4): 395–415.

Hawhee, D. 2004. *Bodily arts: Rhetoric and athletics in ancient Greece.* Austin: University of Texas Press.

Hayne, S. C. 2005. The use of pattern-communication tools and team pattern recognition. *IEEE Transactions on Professional Communication* 48 (4): 377–390.

Hedstrom, M. 2002. Archives, memory, and interfaces with the past. *Archival Science* 2:21–43.

Hovde, M. R. 2001. Research tactics for constructing perceptions of subject matter in organizational contexts: An ethnographic study of technical communicators. *Technical Communication Quarterly* 10 (1): 59–95.

Hutchins, E. 1995a. *Cognition in the wild.* Cambridge, MA: MIT Press.

———. 1995b. How a cockpit remembers its speeds. *Cognitive Science* 19:265–288.

Johnson, G. 1991. *In the palaces of memory: How we build the worlds inside our heads.* New York: Knopf.

Johnson, R. R. 1998. *User-centered technology: A rhetorical theory for computers and other mundane artifacts.* Albany: SUNY Press.

———. 2010. Craft knowledge: Of disciplinarity in writing studies. *College Composition and Communication* 61 (4): 673–690.

Johnson-Eilola, J. 1996. Relocating the value of work: Technical communication in a post-industrial age. *Technical Communication Quarterly* 5(3): 245–270.

Jones, S. L. 2005. From writers to information coordinators: Technology and the changing face of collaboration. *Journal of Business and Technical Communication* 19 (4): 449–467.

Jones, W., & H. Bruce. 2005. *A report on the NSF-sponsored workshop on personal information management,* Seattle, WA, 2005. Retrieved January 15, 2010, from http://pim.ischool.washington.edu/final%20PIM%20report.pdf.

Jones, W., & J. Teevan. 2007. Introduction. In *Personal information management,* edited by W. Jones & J. Teevan, 3–20. Seattle: University of Washington Press.

Kanter, R. M. 1983. The change masters: Innovation for productivity in the American corporation. New York: Simon & Schuster.

Kaptelinin, V., & M. Czerwinski. 2007. Introduction: The desktop metaphor and new uses of technology. In *Beyond the desktop metaphor: Designing integrated digital work environments,* edited by V. Kaptelinin & M. Czerwinski, 1–12. Cambridge, MA: MIT Press.

Kaptelinin, V., & B. A. Nardi. 2006. *Acting with technology: Activity theory and interaction design.* Cambridge, MA: MIT Press.

Kellogg, R. T. 1996. A model of working memory in writing. In *The science of writing: Theories, methods, individual differences, and applications,* edited by C. M. Levy & S. Ransdell, 57–71. Mahwah, NJ: Lawrence Erlbaum Associates.

Kikoski, C. K., & J. F. Kikoski. 2004. *The inquiring organization: Tacit knowledge, conversation, and knowledge creation; Skills for 21st-century organizations.* Westport, CT: Praeger.

Kim, L., & C. Tolley. 2004. Fitting academic programs to workplace marketability: Career paths of five technical communicators. *Technical Communication* 51 (3): 376–386.

Kinneavy, J. L. 1986. *Kairos*: A neglected concept in classical rhetoric. In *Rhetoric and praxis: The contribution of classical rhetoric to practical reasoning*, edited by J. D. Moss, 79–105. Washington, DC: Catholic University of America Press.

Kirsh, D. 1995. The intelligent use of space. *Artificial Intelligence* 73:31–68.

———. 2000. A few thoughts on cognitive overload. *Intellecta, 2000/1* 30:19–51.

Klingberg, T. 2009. *The overflowing brain: Information overload and the limits of working memory.* New York: Oxford UP.

Kress, G., C. Jewitt, J. Ogborn, & C. Tsatsarelis. 2001. *Multimodal teaching and learning: The rhetorics of the science classroom.* London: Continuum.

Krishnan, A., & S. Jones. 2005. TimeSpace: Activity-based temporal visualization of personal information spaces. *Personal and Ubiquitous Computing* 9:46–65.

Krull, R., J. Friauf, J. Brown-Grant, & A. Eaton. 2001. Usability trends in an online help system: User testing on three releases of help for a visual programming language. In *Proceedings of 2001 IEEE IPCC*, 19–26.

Lakoff, G., & M. Johnson. 1999. *Philosophy in the flesh: The embodied mind and its challenge to Western thought.* New York: Basic Books.

Lam, A. 2000. Tacit knowledge, organizational learning and societal institutions: An integrated framework. *Organization Studies* 21 (3): 487–513.

Lane, R. 2013. John Sculley just gave his most detailed account ever of how Steve Jobs got fired at Apple. *Forbes* (September 9). Retrieved December 16, 2013, from http://www .forbes.com/sites/randalllane/2013/09/09/john-sculley-just-gave-his-most-detailed -account-ever-of-how-steve-jobs-got-fired-from-apple/.

Larsen, R. L., & H. D. Wactlar. 2003. *Knowledge lost in information: Report of the NSF workshop on research directions for digital libraries*, NSF Post Digital Library Futures Workshop, Chatham, MA, June 15–17, 2003. Retrieved August 2, 2007, from http://www.sis.pitt.edu/~dlwkshop/report.pdf.

Lave, J. 1991. Situated learning in communities of practice. In *Perspectives on Socially Shared Cognition*, edited by L. Resnick, J. Levine, & S. Teasley, 63–82. Washington, DC: American Psychological Association.

Lave, J., & E. Wenger. 1991. *Situated learning: Legitimate peripheral participation.* New York: Cambridge UP.

Law, J. 2007. Making a mess with method. In *The Sage handbook of social science methodology*, edited by W. Outhwaite & S. P. Turner, 595–606. Beverly Hills, CA: Sage.

Leontiev, A. 1978. *Activity, consciousness, and personality.* Englewood Cliffs, NJ: Prentice-Hall. (Original work published in Russian in 1975.)

Luria, A. R. 1987. *The mind of a mnemonist: A little book about a vast memory.* Translated by L. Solotaroff. Cambridge, MA: Harvard UP.

MajesticSEO. 2013. Retrieved December 16, 2013, from http://www.majesticseo.com/.

Malone, T. W. 1983. How do people organize their desks? Implications for the design of office information systems. *ACM Transactions on Office Information Systems* 1 (1): 99–112.

Marchionini, G. 1995. *Information seeking in electronic environments.* New York: Cambridge UP.

McKenzie, A. 2008. First flight. In *Readings for technical communication*, edited by J. MacLennan, 15–28. Toronto: Oxford UP Canada.

McLean, L. D. 2005. Organizational culture's influence on creativity and innovation:

A review of the literature and implications for human resource development. *Advances in Developing Human Resources* 7 (2): 226–246.

McNeill, D. 1992. *Hand and mind: What gestures reveal about thought.* Chicago: University of Chicago Press.

———. 2005. *Gesture and thought.* Chicago: University of Chicago Press.

Miles, M. B., & A. M. Huberman. 1994. *Qualitative data analysis: An expanded sourcebook.* 2nd ed. Thousand Oaks, CA: Sage Publications.

Moè, A., & R. De Beni. 2005. Stressing the efficacy of the loci method: Oral presentation and the subject-generation of the loci pathway with expository passages. *Applied Cognitive Psychology* 19:95–106. Published online September 16, 2004, in Wiley InterScience (www.interscience.wiley.com).

Mott, R. K., & J. D. Ford. 2007. The convergence of technical communication and information architecture: Managing single source objects for contemporary media. *Technical Communication* 54 (1): 27–45.

Murphy, J. J. 2001. The key role of habit in Roman writing instruction. In *A short history of writing instruction,* 2nd ed., edited by J. J. Murphy, 35–78. Mahwah, NJ: Lawrence Erlbaum Associates.

———. 2002. The metarhetoric of Aristotle, with some examples from his *On memory and recollection. Rhetoric Review* 21 (3): 213–228.

Nader, K. 2003. Memory traces unbound. *Trends in Neurosciences* 26 (2): 65–72.

Nonaka, I., & N. Konno. 1998. The concept of "Ba": Building a foundation for knowledge creation. *California Management Review* 40 (3): 40–54.

Norman, D. A. 2002. *The design of everyday things.* New York: Basic Books.

Ockelford, A. 2007. *In the key of genius: The extraordinary life of Derek Paravicini.* London: Arrow.

Ong, W. J. 1982. *Orality and literacy: The technologizing of the word.* New York: Routledge.

Orlikowski, W. J., & J. Yates. 1994. Genre repertoire: The structuring of communicative practices in organizations. *Administrative Science Quarterly* 39:541–574.

Orr, J. E. 1996. *Talking about machines: An ethnography of a modern job.* Ithaca, NY: ILR Press.

Oxford English dictionary. 1989. 2nd ed. Oxford: Oxford UP.

Perry, M. J., R. Fruchter, & D. Rosenberg. 1999. Co-ordinating distributed knowledge: A study into the use of an organizational memory. *Cognition, Technology, and Work* 1:142–152.

Poster, W. R., & S. Prasad. 2005. Work-family relations in transnational perspective: A view from high-tech firms in India and the United States. *Social Problems* 52 (1): 122–146.

Pratt, J. A. 1998. Where is the instruction in online help systems? *Technical Communication* 45 (1): 33–37.

Prior, P. 2004. Tracing process: How texts come into being. In *What writing does and how it does it: An introduction to analyzing texts and textual practices,* edited by C. Bazerman & P. Prior, 167–200. Mahwah, NJ: Lawrence Erlbaum Associates.

Prior, P., & J. Shipka. 2003. Chronotopic lamination: Tracing the contours of literate activity. In *Writing selves/writing societies: Research from activity perspectives,* edited by C. Bazerman & D. R. Russell. Fort Collins, CO: WAC Clearinghouse and Mind, Culture, and Activity. Retrieved March 24, 2006, from http://wac.colostate.edu/books/selves_societies/.

Ravasio, P., & V. Tscherter. 2007. Users' theories of the desktop metaphor, or why we should seek metaphor-free interfaces. In *Beyond the desktop metaphor: Designing integrated digital work environments*, edited by V. Kaptelinin & M. Czerwinski, 265–294. Cambridge, MA: MIT Press.

Reich, R. B. 1991. *The work of nations: Preparing ourselves for 21st-century capitalism.* New York: Vintage Books.

Reynolds, R. E., G. M. Sinatra, & T. L. Jetton. 1996. Views of knowledge acquisition and representation: A continuum from experience centered to mind centered. *Educational Psychologist* 31 (2): 93–104.

Rice, R. E., M. McCreadie, & S.-J. Chang. 2001. *Accessing and browsing information and communication.* Cambridge, MA: MIT Press.

Rockley, A., P. Kostur, & S. Manning. 2003. *Managing enterprise content: A unified content strategy.* Boston: New Riders.

Salvo, M. J. 2001. Ethics of engagement: User-centered design and rhetorical methodology. *Technical Communication Quarterly* 10 (3): 273–290.

Saugstad, T. 2002. Educational theory and practice in an Aristotelian perspective. *Scandinavian Journal of Educational Research* 46 (4): 373–390.

Schwaber, K., & M. Beedle. 2002. *Agile software development with scrum.* Upper Saddle River, NJ: Prentice Hall.

Sellen, A. J., & H. R. Harper. 2002. *The myth of the paperless office.* Cambridge, MA: MIT Press.

Sharples, M. 1996. An account of writing as creative design. In *The science of writing: Theories, methods, individual differences, and applications*, edited by C. M. Levy & S. Ransdell, 127–148. Mahwah, NJ: Lawrence Erlbaum Associates.

Shotter, J. 1990. The social construction of remembering and forgetting. In *Collective remembering*, edited by D. Middleton & D. Edwards, 120–138. London: Sage.

Slack, J. D., D. J. Miller, & J. Doak. 2004. The technical communicator as author: Meaning, power, authority. In *Central works in technical communication*, edited by J. Johnson-Eilola & S. A. Selber, 160–174. New York: Oxford UP. (Original work published 1993.)

Slack, J. D., & J. M. Wise. 2005. *Culture + technology: A primer.* New York: Peter Lang.

Slattery, S. 2007. Undistributing work through writing: How technical writers manage texts in complex information environments. *Technical Communication Quarterly* 16 (3): 311–325.

Small, J. P. 1997. *Wax tablets of the mind: Cognitive studies of memory and literacy in classical antiquity.* New York: Routledge.

Sorabji, R. 1972. Introduction. In *Aristotle on Memory*, translated by R. Sorabji, 1–10. Providence, RI: Brown UP.

Spender, J. C. 1996. Organizational knowledge, learning and memory: Three concepts in search of a theory. *Journal of Organizational Change Management* 9 (1): 63–78. Retrieved April 19, 2012, from ABI/INFORM Global (document ID: 117542613).

Spinelli, G., M. Perry, & K. O'Hara. 2005. Understanding complex cognitive systems: The role of space in the organization of collaborative work. *Cognition, Technology, and Work* 7:111–118.

Spinuzzi, C. 2003a. Compound mediation in software development: Using genre ecologies to study textual artifacts. In *Writing selves/writing societies: Research from activity perspectives*, edited by C. Bazerman & D. R. Russell. Fort Collins, CO: WAC Clearinghouse and Mind, Culture, and Activity. Retrieved January 10, 2008, from http://wac.colostate.edu/books/selves_societies/.

———. 2003b. *Tracing genres through organizations: A sociocultural approach to information design*. Cambridge, MA: MIT Press.

———. 2008. *Network: Theorizing knowledge work in telecommunications*. New York: Cambridge UP.

Star, S. L. 1999. The ethnography of infrastructure. *American Behavioral Scientist* 43 (3): 377–391.

Strauss, A. 1987. *Qualitative analysis for social scientists*. New York: Cambridge UP.

Sullivan, D. L., M. S. Martin, & E. R. Anderson. 2003. Moving from the periphery: Conceptions of ethos, reputation, and identity for the technical communicators. In *Power and legitimacy in technical communication*, vol. 1, *The historical and contemporary struggle for professional status*, edited by T. Kynell-Hunt & G. J. Savage, 115–136. Amityville, NY: Baywood.

Sullivan, P., & J. E. Porter. 1997. *Opening spaces: Writing technologies and critical research practices*. Westport, CT: Ablex Publishing Corp.

Swarts, J. 2004. Cooperative writing: Achieving coordination together and apart. *Proceedings of the 2004 ACM Conference on Documentation*, 83–89.

Tedlock, B. 2003. Ethnography and ethnographic representation. In *Strategies of qualitative inquiry*, 2nd ed., edited by N. K. Denzin & Y. S. Lincoln, 165–213. Thousand Oaks, CA: Sage Publications.

Thayer, S. M., & P. Steenkiste. 2003. An architecture for the integration of physical and informational spaces. *Personal and Ubiquitous Computing* 7:82–90.

Tomalin, C. 2002. *Samuel Pepys: The unequalled self*. New York: Knopf.

U.S. Department of Labor, Bureau of Labor Statistics. 2011. *Women in the labor force: A databook*. Retrieved August 2, 2012, from http://www.bls.gov/cps/wlf-databook-2011.pdf.

Vickers, D., & S. Fox. 2010. Towards practice-based studies of HRM: An actor-network and communities of practice informed approach. *International Journal Of Human Resource Management* 21 (6): 899–914.

Vygotsky, L. 1978. *Mind in society: The development of higher psychological processes*. Cambridge, MA: Harvard UP.

Wahl, S. 2003. Learning at work: The role of technical communication in organizational learning. *Technical Communication* 50 (2): 247–258.

Walsh, J. P., & G. R. Ungson. 1991. Organizational memory. *Academy of Management Review* 16 (1): 57–91.

Wenger, E. 1998. *Communities of practice: Learning, meaning, and identity*. New York: Cambridge UP.

Wenger, E., R. McDermott, & W. M. Snyder. 2002. *Cultivating communities of practice: A guide to managing knowledge*. Boston: Harvard Business School Press.

Whittemore, S. 2008. Metadata and memory: Lessons from the canon of *memoria* for the design of content management systems. *Technical Communication Quarterly* 17 (1): 88–109.

Wick, C. 2000. Knowledge management and leadership opportunities for technical communicators. *Technical Communication* 47 (4): 515–529.

Winsor, D. A. 1996. *Writing like an engineer: A rhetorical education.* Mahwah, NJ: Lawrence Erlbaum.

———. 2001. Learning to do knowledge work in systems of distributed cognition. *Journal of Business and Technical Communication* 15 (1): 5–28.

———. 2003. *Writing power: Communication in an engineering center.* Albany: State University of New York Press.

Wurman, R. S. 2001. *Information anxiety 2.* Indianapolis: QUE.

Yates, J. 1989. *Control through communication: The rise of system in American management.* Baltimore: Johns Hopkins UP.

Zhang, J., & G. Marchionini. 2005. Evaluation and evolution of a browse and search interface: Relation browser++. In *Proceedings of the 2005 National Conference on Digital Government Research*, Atlanta, 179–188.

Zuboff, S. 1988. *In the age of the smart machine: The future of work and power.* New York: Basic Books.

INDEX

Page numbers in italics refer to figures.

Gabriel, Y., 51, 166
Gee, J. P., 80, 211
Geisler, C., 120
Gene, 87–88, 107, 113
gesturing practices, 51; of Angela, 170, 171,
172, 179–82; and Aristotle's phantasm,
40; and articulation work, 38; difficulties
in studying, 179; as "filling in," 181; in
information retrieval, 20; of Lucy, 123,
153–55, 221n4; in Robert's meeting with
Andie and Gene, 87; in Robert's meeting
with Wallace, 91; as spontaneous, 179;
"tag" pieces of information, 205; Wenger
on, 57
Goodenough, W. H., 133
Goodwin, C., 45
Goodwin, M. H., 45
grounded theory, 48
groupware, 61, 71

habit: Aristotle on, 43–44; changing to fit
altered circumstances, 34, 36; embodied,
35; habitual browsing, 151–52; in novice's transformation into master, 79; in
phronesis knowledge, 33, 93; repetition
in, 130, 147; in Roman education, 41; in
situated cognition, 42–43; transforming
information in knowledge through, 130.
See also *hexis*
Hart, H., 10–11
Hawhee, D., 32, 33, 35–36, 37, 43, 194
Hayes, J. R., 53
hexis: of Angela, 175, 193, 194, 198; as barrier,
36; as bridge, 35; embodied, 33, 35–36,
109, 137, 194, 198; as key concept of
rhetorical memory, 29, 33–36; Lucy seeks
to achieve, 131, 133, 137, 139, 143, 144,
156, 157, 200; as mastery, 33, 43, 147;
multiple, 36, 194; in novice's transformation into master, 79; and *phronesis*, 33,
34, 35, 93, 131, 192; and reminding, 107;
repetition in, 147; of Robert, 107, 109,
200; and situated cognition, 43, 44; at
Software Unlimited, 73. See also habit
Hovde, M. R., 11–12
Hull, G., 211
Hutchins, Edwin, 15, 21, 44

improvisation, 1, 55, 60, 62–64, 65–66
influence, bidirectionality of, 64–66, 71
information: anxiety regarding, 7; background, 162; in content "silos," 12, 71;

185, 203; cultural factors in storing and
retrieving, 20; fragmentation of, 9, 11,
72; islands of, 9; memory regimes cue
and constrain circulation of, 201–3;
offloading, 18, 121–22; outdated, 9, 49;
in packets, 27; perspectives on, 13–19;
physical, embodied components in storing and retrieving, 20–21; pushed versus
pulled, 6–8; raw, 8, 14, 128, 156; relevant,
8, 17, 112; sharing, 13, 18, 66, 72, 164,
199; sorting through, 4, 6, 8; stored,
11, 16, 19, 20, 26, 27, 32, 38, 52, 53, 55,
115, 147, 204; strategic reserve of, 102;
transforming into knowledge, 19, 20, 22,
26, 28, 61, 106, 128–29, 130–31, 152–53,
155–56, 157, 166, 173, 196, 200–201, 203,
204; where critical information resides
and how it moves through the organization, 22, 93, 196, 204
Information Development Department:
adversarial relationship with other team
members, 111; Agile development methodology at, 73–75; archiving by, 104–5,
110–11; browsing by, 150; as community
of practice, 56; critical importance of,
203; Daily Scrums of, 75, 178; difficulty
acquiring information about company's
products and their development history,
108, 117; documentation leads, 80–81;
information overload of, 205; infrastructure, workspaces, and tools of, 209–11,
210; knowledge about users difficult
to acquire by, 113–15; members paired
with training department members, 168;
origin of, 190; participants in this study,
77, 208–9; "Petri Dish" area for, 168, 169,
176, 209–10, 210; and *phronesis*, 93–94,
108, 114–15, 116; software engineers'
goodwill required by, 94; user advocacy
responsibility of, 76, 94, 108, 111, 117.
See also Angela; Becky; Lucy; Monica;
Peter; Robert
information management: memory politics
as confounding aspect of, 21; memory
practices and, 20; memory regimes and,
25, 48–51; overview of, 1–22; personal
information management, 17–18, 49;
as rhetorical memory practices, 19–22,
34, 117; technical communicators as
information managers, 12–13, 18, 66;
technology in, 2
information overload: browsing as cause